Natural Disasters and Climate Change

Stéphane Hallegatte

# Natural Disasters and Climate Change

## An Economic Perspective

Springer

Stéphane Hallegatte
Sustainable Development Network
World Bank
Washington, DC, USA

ISBN 978-3-319-08932-4 ISBN 978-3-319-08933-1 (eBook)
DOI 10.1007/978-3-319-08933-1
Springer Cham Heidelberg New York Dordrecht London

Library of Congress Control Number: 2014949246



*Cover image*: Close-up view of the eye of Hurricane Isabel taken by one of the Expedition 7 crewmembers onboard the International Space Station (ISS). Image provided by NASA.

Printed on acid-free paper

Springer is part of Springer Science+Business Media (www.springer.com)

# Prologue

New Orleans was founded in 1718 and by the early nineteenth century had grown into the largest city in the southern United States. Its land area protected by natural levees was very small, so as the city expanded, it spread into marshland that was drained using pumps, drainage canals, and artificial levees. More reliable electric pumps and the development of better levees at the start of the twentieth century allowed for accelerated development. In the years since, however, weather-related catastrophes have become common in and around the city of New Orleans.

In 1915, a hurricane overflowed the protection system along the city's Lake Pontchartrain shore. Water levels reached 4 m in some districts, and it took 4 days to pump the water from the city. The government responded by upgrading pump stations and raising levees along the drainage canals. In 1947, another hurricane hit the city, and the levees failed again. Thirty square miles flooded, and 15,000 people had to be evacuated. Again, major improvements to the protection system followed in the immediate aftermath of the disaster, with levees being raised and extended.

In 1965, Hurricane Betsy made landfall, and New Orleans flooded again. About 13,000 homes filled with water, leaving 60,000 people homeless and causing 53 deaths and more than $1 billion in damage. This led to the passing of the Flood Control Act of 1965 by the U.S. Congress and to an ambitious plan to protect New Orleans. The plan was to be fully implemented within 13 years, but in the face of numerous difficulties, including conflicts with environmental protection movements, it remained stalled for about two decades. It was eventually revised into the "high level plan." The implementation of that plan was 60–90 % complete when Hurricane Katrina struck in 2005, leading to the flooding of 80 % of the city and unprecedented human and economic damages. The complete failure of the protection system in 2005 demonstrated that both construction and maintenance had not been adequately supervised and monitored.

Over the past 100 years and four disasters, the New Orleans region has experienced large socioeconomic and environmental changes. In particular, the local sea level rose by 5 cm per decade, about 50 cm in all, because of geological factors:

the soil was (and still is) sinking, a process referred to as "subsidence." Failure to protect the New Orleans population from disasters, illustrated by a decrease in the city's population since its peak in 1965, can provide important lessons regarding how to manage risks in other locations.

Indeed, the global sea level rise due to climate change will affect all coastal cities in the future, and this rise is expected to be of the same order of magnitude as what was experienced in New Orleans in the last 100 years. Over the coming decades, many cities around the world will thus experience the same changes in risk as New Orleans did in the past, and one can hope they do not have to go through a similar series of disasters.

Fortunately, risk management also offers more positive stories. In the Netherlands, subsidence also made the local sea level rise by about 2 cm per decade during the twentieth century. A flood there in 1953 caused more than 1,800 deaths and extensive damage. The response to this event went beyond just engineering more and better protection. The Delta committee was created to manage the response from institutional, legal, and technical perspectives. In 1960, the committee published the Delta Plan, which included an engineering section, the Delta Works, but also a new approach to the management of flood risks. The Delta committee determined an acceptable level of flood risk in different regions of the country through a combination of economic analyses and political decisions. From there, it derived an optimum level of protection, which could then be used by engineers to design protection systems.

Risk management in the Netherlands does not exhibit the same cycle as in New Orleans, where defense improvements have been driven by disasters demonstrating the weakness of protections. The Dutch Law on Water Defences requires that water levels and wave heights used in risk analyses and in the design of protections be updated every 5 years and that water defenses be evaluated for these new conditions. Such a response does not reduce risk to zero, and the Netherlands dealt with flooding again in the 1990s. But the 5-year updates ensure that changing demographic, economic, and environmental conditions are taken into account in the design, maintenance, and upgrades of flood defenses, even if no disaster has occurred.

New Orleans' history shows how socioeconomic and environmental changes can increase both the risk and the damage when storms strike. The Netherlands example suggests that good risk management can reduce the losses. With the right policies and decisions, future risks can be managed, even as climate change increases vulnerability in some places. Strengthening risk management will not eliminate disasters, but it will avoid many crises, save lives, and reduce losses and suffering.

We cannot predict how well we will be able to manage future risks in the face of climate change, but much can be done to increase the odds of a scenario in which ever-changing socioeconomic and environmental conditions are accounted for, disaster risks are reduced as much as possible, affected populations are supported in post-disaster situations, and climate change impacts are as limited as possible.

This book provides insights into how to manage natural risks in a changing environment. Many remarkable books investigate the social, health, and psychological aspects of catastrophes. This book tries to complement them by taking an economist's point of view and providing economic tools to inform policymakers for taking better decisions regarding risk management so we can prevent the avoidable catastrophes and cope with the unavoidable ones.

# Acknowledgments

This book is based on articles published between 2007 and 2013, with many coauthors and collaborators. My main collaborators on this work are Philippe Ambrosi, Jan Corfee-Morlot, Patrice Dumas, Michael Ghil, Susan Hanson, Fanny Henriet, Jean-Charles Hourcade, Robert Lempert, Olivier Mestre, Nicolas Naville, Robert Nicholls, Valentin Przyluski, Nicola Ranger, Ankur Shah, Lionel Tabourier, and Vincent Viguié.

During these years, hundreds (thousands?) of hours of discussion with Patrice Dumas played a critical role in shaping my ideas. My modeling work also benefited immensely from the support and insights provided by Jean-Yves Grandpeix and Alain Lahellec from the Laboratoire de Météorologie Dynamique and by the TEF-ZOOM modeling community. The book was completed while I was part of the core writing team of the 2014 World Development Report, entitled "Risk and Opportunity," and the entire team provided very influential ideas, especially about the process of risk management and the institutional side of the issues.

Other friends and colleagues have to be acknowledged for the exchanges we had over these years and the support they have provided to me: Paolo Avner, Anthony Bigio, Auguste Boissonnade, Laurens Bouwer, Jean-Louis Dufresne, Kris Ebi, Ottmar Edenhofer, Kerry Emanuel, Sam Fankhauser, Chris Field, Francis Ghesquière, Colin Greene, Goeffrey Heal, Jean Jouzel, Nidhi Kalra, Howard Kunreuther, Norman Loayza, Reinhart Mechler, Erwann Michel-Kerjan, Robert Muir-Wood, Roger Pielke Jr., Julie Rozenberg, Reimund Schwarze, Eric Strobl, Richard Tol, Vincent Viguié, Adrien Vogt-Schilb, and Gary Yohe. Several benevolent (and pitiless) reviewers helped improve the manuscript, including Laura Bonzanigo, Fabrice Chauvin, Jun Rentschler, Julie Rozenberg, Vincent Viguié, and Adrien Vogt-Schilb. They offered very important suggestions to improve the organization and content of the manuscript and provided critical feedback on the framework used in this book. Stacy Morford kindly edited the prologue and introduction.

Most of the work presented in this book has been done in the context of my research at the Centre International de Recherche sur l'Environnement et le

Développement (CIRED), directed by Jean-Charles Hourcade. CIRED offered a great environment for academic research, through frequent and invaluable informal interactions and discussions, on the topic of this book and on many others. I thank the entire team for the ideas that were shared during my time there.

From 2007 to 2012, I was a researcher for Météo-France and professor at the Ecole Nationale de la Météorologie, led by François Lalaurette. During that time and beyond, I benefited from the continuous support of Alain Ratier, deputy director of Météo-France. Additional financial support was provided by the European Commission through multiple projects (E2-C2, Ensembles, Weather, ConHaz), by Risk Management Solutions, and by the OECD through the leadership of Jan Corfee-Morlot.

The book was completed while I was at the World Bank, in the office of the Sustainable Development Network Chief Economist, Marianne Fay. Her friendly support and advice have been critical for finding the time and energy to complete this project.

# Contents

# List of Figures

# List of Tables

# Chapter 1
# Introduction and Summary

**Abstract** This chapter introduces the book by summarizing three major debates regarding the link between disaster risk management and climate change adaptation, namely the existence and magnitude of long-term economic impacts of natural disasters, the idea that development is the solution to excessive natural risks and climate change adaptation challenge, and the link between trends in natural disasters and climate change. The chapter also presents the main messages of the book, stressing the existence of large potential gains and synergies from a combination of disaster-risk-reduction, resilience-building and adaptation policies. It stresses that disaster risk management policies need to shift from a purely negative stance and follow a more positive and holistic approach that is fully integrated in development planning and aims at a more resilient development.

**Keywords** Disaster risk management • Climate change adaptation • Resilience • Development

The amount of economic damage caused by natural disasters has increased over the past three decades (Fig. 1.1). Moreover, a few recent natural disasters have changed our perception of socioeconomic vulnerability. The 2005 landfall of Katrina in New Orleans showed that large-scale destruction and large losses of lives are not limited to developing countries, and that one event can lead to a local economic collapse. The 2004 tsunami in Asia demonstrated that many countries can be affected by a single exceptional event. The 2010 earthquake in Haiti showed how an entire country can be paralyzed by a disaster, making recovery and reconstruction extremely challenging. The 2010 fires in Russia illustrated how a disaster in one country can have global consequences through commodity markets and food prices. Further, the Icelandic volcano Eyjafjallajökull forced the cancellation of thousands of flights, showing that, even in the absence of destruction, a hazard can create heavy perturbations to the functioning of the global economic system. The earthquake in Japan in March 2011 showed that even the best-prepared regions can be overwhelmed by an exceptionally intense event, and that global supply chains can be heavily affected by a disaster. And the landfall of Superstorm Sandy in 2012 in New York City proved that even the richest cities are sometimes under-prepared for large-scale weather events.

 
S. Hallegatte, *Natural Disasters and Climate Change*,
DOI 10.1007/978-3-319-08933-1_1

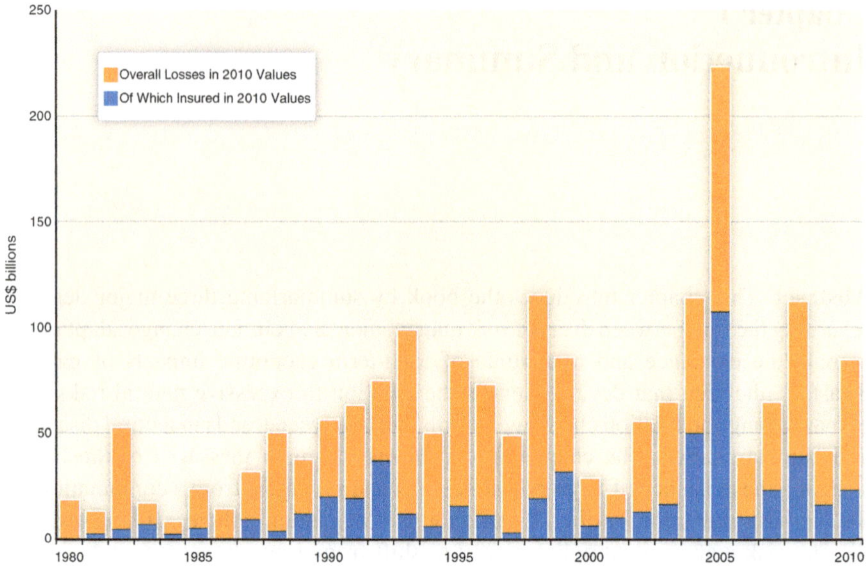


**Fig. 1.1** The overall losses and insured losses from weather- and climate-related disasters worldwide (in 2010 US$). These data for weather- and climate-related 'great' and 'devastating' natural catastrophes are plotted without inclusion of losses from geophysical events. A catastrophe in this data set is considered 'great' if the number of fatalities exceeds 2,000, the number of homeless exceeds 200,000, the country's GDP is severely hit, and/or the country is dependent on international aid. A catastrophe is considered 'devastating' if the number of fatalities exceeds 500 and/or the overall loss exceeds US$650 million (in 2010 values) (Data from Munich Re (2011). Source: IPCC (2012))

Concerns about future vulnerability to disasters have been raised repeatedly in recent years, in both the scientific and policy communities. Decision-makers and policy leaders promise resolute action after each large event. But beyond recognizing the need for action, little has been said about what should be done and what can be done. This book investigates these questions to help design policy responses.

To do so, a better understanding of disaster consequences is urgently needed, and three main scientific debates must be resolved.

The first debate concerns the economic impact of natural disasters on development. Beyond the human toll and the unquestionable immediate impact on welfare, should we be really concerned about long-term economic consequences of disasters? Albala-Bertrand claimed in 1993 that disasters were a problem of development, but not a problem for development. In other terms, he estimated that disasters are not a macroeconomic threat, and that their long-term impacts are generally overestimated. This view has been supported by a few scholars (e.g., Skidmore and Toya 2002, 2007), and challenged by many others (e.g., Benson and Clay 2004; Noy and Nualsri 2007; Noy 2009; Hochrainer 2009; Jaramillo 2009; Raddatz 2009; Strobl 2011; Felbermayr and Gröschl 2013). The recent increase in economic losses from disasters has weighed into the debate, and natural disasters and disaster risk reduction are now considered serious economic issues. In 2005, the Hyogo Framework for Action was adopted by the international community to coordinate and amplify efforts to reduce disaster consequences. But in spite of important policy statements and the huge impact of disasters on populations, knowledge is surprisingly weak on how disasters influence growth, development, and poverty reduction. In particular, there is still no consensus on the effect of disasters on long-term development.

Second, there have been heated discussions on whether development is the solution to increasing disaster risk. According to many scholars (e.g., Skidmore and Toya 2007; Mendelsohn et al. 2012; Bakkensen 2013; Felbermayr and Gröschl 2013), disaster losses decrease – at least in relative economic terms – with wealth, and economic development will solve the disaster risk issue. Others have noticed that development can be either risk-decreasing (e.g., building of better coastal protection infrastructure) or risk-increasing (e.g., because urbanization can lead to the development of slums in high-risk areas), and that rich countries also have problems dealing with natural hazards (World Bank 2009, 2010; Hallegatte 2013). The landfall of Superstorm Sandy in New York City has reinforced the idea that richer areas are not always less vulnerable, and the ranking of the most vulnerable cities proposed by Hallegatte et al. (2013) shows that higher income does not always translate into better protection and less exposure to hazards. Here, the question is whether development policies and projects need to account explicitly for disaster risks and if trade-offs are needed or desirable between economic growth and disaster risk reduction.

A last debate concerns the connection between natural disasters and climate change, an issue often raised after disasters. It is indeed difficult to distinguish long-term trends – such as a climate change signal – from natural variability and the chaotic dynamics of the climate system. Progress has been achieved in attributing physical hazards to climate change, with statements regarding the physical hazard probability (e.g., heat waves similar to the one that affected the American Midwest and Northeast in July 2012 are now four times as likely because of climate change; see Peterson et al. 2013). But it is more difficult to analyze disaster impacts and losses and their trends. Disasters are too rare to allow for clear-cut conclusions from rigorous statistical analysis. Moreover, natural hazards affect societies and economies that are in permanent evolution, and a large fraction of the increase in disaster losses arises from the growth of population and assets in at-risk areas. That makes it even more difficult to identify and attribute a possible residual trend in losses to climate change or any other environmental change.

Concerns about the role of climate change in the observed increase in disaster losses have nevertheless helped raise awareness of disaster consequences. Disaster risk reduction initiatives (e.g., the Hyogo Framework for Action, the Global Facility for Disaster Risk Reduction and Recovery) and climate change adaptation actions (e.g., in the United Nations Framework Convention on Climate Change, UNFCCC) are now more closely related. For instance, UNFCCC discussions now include debates on insurance and risk prevention funding and a work program on "loss and damages." The risk management community has also been increasingly interested in how global warming changes the best approaches to risk analysis. In the scientific community, this link between climate change and disaster risk is best exemplified by the "SREX" Report (the Special Report of the Intergovernmental Panel on Climate Change (IPCC) entitled "Managing the Risks of Extreme Events and Disasters to Advance Climate Change Adaptation"). Current research concludes that past increases in disaster losses cannot be attributed to climate change, but it does not mean that this will remain true in the future: climate change will affect more and more weather patterns in the next decades, changing how risk management strategies should be designed.

This book summarizes recent investigations related to these questions, and provides insights to guide policy action in the risk reduction and climate change domains.

**The first part of this book (Chap. 2) focuses on the economic consequences of a disaster and discusses the definition of its "economic cost." It stresses that a natural disaster is not a natural event, but the combination of a natural hazard (e.g., a hurricane) with a human system that is exposed to it and suffers from damages and perturbations.** It reviews concepts such as the economic cost of a disaster, including direct and indirect losses, market and non-market losses, and consumption and output losses. It describes some of the most important mechanisms that determine the economic consequences of a disaster, such as the response of prices or the propagation of impacts through supply chains. It also discusses the various tools that have been developed to measure and assess output losses from disasters, covering econometric analyses, input-output and computable general equilibrium models. This discussion on the economic consequences of disasters concludes with a definition of a concept of economic resilience.

**Part II goes from the consequences of one disaster to the notion of risk, defined as the combination of a potential natural hazard (e.g., an earthquake with a given probability of occurrence every year), the physical exposure of a human system (e.g., a city that sits on a fault line subject to the earthquake), the sensitivity to the hazard (e.g., buildings that would be damaged if the earthquake occurred), and the resilience of the affected human system (e.g., its ability to recover and reconstruct).** Chapter 3 proposes a brief overview of natural risks today and discusses current trends in disaster losses by introducing the notion of risk-taking and the benefits from risk-taking. Taking into account the benefits of taking risks helps us to understand why many people and assets are located in at-risk areas, such as flood plains and coastal zones. It also helps introduce potential policy options to reduce risk. Chapter 4 reviews how climate change may affect natural hazards in the future, stressing the heterogeneity in its effect on hazard depending on the type of hazard (e.g., heavy rainfall vs. snow storm) and the region of the world. Chapter 5 proposes a methodology to assess the impact of climate change on natural hazards and disaster losses, and applies it to three case studies: hurricanes in the United States; storm surges in Copenhagen; and heavy precipitation in Mumbai. These three case studies are chosen to illustrate different aspects of the questions and the different tools that can be mobilized to provide inputs and information to decision-makers.

**Part III makes a link between natural disasters and climate change adaptation and discusses how the two can be managed together. It shows that climate change adaptation and disaster risk management have to be designed in a consistent and unified framework and that synergies exist between reducing current risk and reducing future climate change vulnerability.** Chapter 6 starts by comparing reactive risk management practices and the proactive approaches that are required in a climate change context, using New Orleans and the Netherlands as case studies. It then presents a cost-benefit analysis approach, with two case studies on New Orleans coastal protections and on investments in early warning systems in developing countries. It stresses the limits of this approach, especially when uncertainty is very large. Chapter 7 highlights strategies that are best able to cope with the large uncertainty (and disagreement) that surround the climate change impacts on natural hazards and disasters. It introduces the robust decision-making approach and suggests flexible and reversible strategies that are best suited to the current situation of high uncertainty on future climates.

This book proposes a few key messages, which can help in understanding and managing disasters.

**First, a natural disaster is not a natural event: it is the combination of a natural hazard and the exposure and vulnerability of a human system.** Risk is often presented as the product of three factors: (1) the hazard, which is the natural event, such as a storm or an earthquake; (2) the exposure, which is the population and assets potentially affected by hazards; and (3) the sensitivity, i.e. the human and economic losses if population and assets are affected by a hazard. With no exposure (nobody and nothing in the area potentially affected), there is no disaster. With no sensitivity (no damage to the exposed population and assets if a hazard occurs), there is no disaster. There is risk and there can be a disaster only if there is a human system that is exposed and sensitive to the natural event.

**Second, there is no single definition of the economic cost of a disaster, and the relevant definition depends on the purpose of the assessment** (e.g., the cost-benefit analysis of prevention measures or the assessment of post-disaster assistance needs). Using welfare metrics, the total cost of natural disasters can be much higher than direct losses, and "indirect losses" play an important role. **Indirect losses (also labeled "higher-order losses" in** Rose 2004**) include all losses that are not provoked by the disaster itself, but by its consequences; they are spanning over a longer period of time than the event, and they affect a larger spatial scale or different economic sectors.** They include several categories of losses, such as lost output due to capital damages (including business interruptions and supply-chain disruptions), macro-economic feedbacks and long-term adverse consequences on economic growth, long term consequences of health effects and social network disruption, and sometimes even the impact on security, cohesion, and political stability. **Indirect losses are found to increase non-linearly with the amount of direct losses: while small events can be measured in an acceptable manner by their direct consequences only, it is impossible to assess the consequences of large-scale disasters without considering their indirect impacts.** Models are still unable to provide reliable estimates of indirect losses, but they help understand the mechanisms and identify policies and measures to reduce them.

**Third, reducing disaster welfare impacts can be done by reducing direct losses (e.g., by improving building norms) or by reducing indirect losses (e.g., by helping households to rebuild). Reducing direct losses can be viewed as increasing the "robustness" of the economic system, while reducing indirect losses can be understood as increasing its "resilience."** The total economic impact of a disaster depends on complex inter-linkages within the economic system, and reducing it is possible through many actions: by relaxing the limits to reconstruction capacity linked to financial, technical, or institutional constraints; by affecting the structure of the economic system, including firm-to-firm network characteristics and inventory management; by improving the interaction with the "rest of the world" and facilitating the fluxes of goods, workers, and funds from outside the affected region.

**Fourth, the existence of a risk can have negative consequences on development, growth, and poverty reduction, even if the risk does not materialize into a disaster.** If households and companies know that they may lose their productive assets at any time because of a disaster, they might decide not to accumulate assets and to consume their income, or to keep their savings in less productive forms. This would have a negative impact on development and economic growth. And because the poorest are more exposed, they are the ones most likely to forgo risky investments that can help them escape poverty, such as taking a loan to use fertilizers in farming. Reciprocally, banning risk-taking would also affect negatively development prospects, since capturing growth opportunities often implies to take new risks (World Bank 2013; Hallegatte 2013). **Risk management is thus not only an instrument to reduce losses from disasters. It is also an instrument to make it possible for individuals and companies to take risks when benefits exceed cost and, thus, to capture opportunities. As such, risk management can be considered an "economic growth instrument" that can help individuals and companies take the risks that are needed to grow their income and wealth and improve their well-being.**

**Fifth, the recent growth in disaster economic losses is overwhelmingly due to socioeconomic trends, but climate change will play an increasing role over time.** Today, growing population and wealth and shifting localization choices toward high-risk areas participate in the increase in disaster losses. But even if current trends in disaster losses can be fully explained by socioeconomic trends, that does not mean that climate change will not play a significant role in the coming decades. Climate change will not amplify all hazards in all regions, but it is very likely to increase some of them in many regions. **Analyses suggest that climate change can increase dramatically natural risks in some contexts, especially where protections have been designed for a stable climate and may quickly become ill-adapted to new conditions.** More than average losses, the increase in the likelihood of the more intense hazards is a serious reason for concern.

**There is no reason natural disaster losses will stop increasing rapidly in the future unless specific risk-reducing actions are implemented.** Development does not automatically reduce natural disaster risks, and natural hazards need to be taken into account in development plans. For instance, current trends toward higher efficiency and productivity (e.g., smaller inventories, fewer suppliers, and just-in-time production) may lead to an increase in the vulnerability of the economic system to the most extreme events. Coastal and flood-plain urbanization in search of agglomeration externalities, such as larger job markets and cheaper transportation, is increasing exposure to floods in many countries. These trends toward higher risks need to be mitigated and counter-balanced by risk-reducing policies and actions, such as better building standards and zoning policies.

**Fortunately, there are many options to reduce future disaster losses, especially through the integration of climate change adaptation and disaster risk management and the implementation of decision-making methodologies able to cope with increased uncertainty.** Stronger hazards due to climate change do not need to translate into more or larger disasters, as proactive risk analysis and management approaches can be used to cope with this trend. These approaches are more difficult to implement than reactive ones and rely on strong institutions and legal frameworks. Since there is no silver bullet policy against risk, designing a policy mix, from physical protection to land-use planning and insurance, is necessary. Moreover, "no-regret" strategies are available, which pay for themselves and are able to reduce future disaster losses at a negative (or limited) cost. However, the uncertainty in future climate conditions means that new decision-making approaches need to be mobilized. Cost-benefit analysis is an important and useful tool, but its limits need to be acknowledged and managed. Alternative methodologies, such as robust decision-making and multi-criteria analysis, can also be used to improve risk management, and they are better able to manage the current context of large uncertainty and disagreement on how future climate hazards will change.

**Finally, risk management policies need to recognize the benefits from risk-taking and avoid suppressing it entirely.** In coastal areas with low transportation costs, international trade can generate additional economic growth, but only if development is allowed in spite of the larger exposure of these zones to flooding. Risk-taking can also increase production when additional investments in urban flood zones generate agglomeration externalities. Disaster risk management policies need to shift from a purely negative stance – indicating where development is prohibited – to a more positive approach, indicating where investments should be directed to reduce risk levels. Risk management should not focus only on at-risk areas, but follow a more holistic approach, one that is fully integrated in development planning.

## References

Bakkensen LA (2013) Adaptation and natural disasters: evidence from global tropical cyclone damages and fatalities. In: EAERE 2013 annual conference, Toulouse

Benson C, Clay E (2004) Understanding the economic and financial impact of natural disasters. The international bank for reconstruction and development. The World Bank, Washington, DC

Felbermayr GJ, Gröschl J (2013) Naturally negative: the growth effects of natural disasters, SSRN Scholarly Paper No. ID 2348054. Social Science Research Network, Rochester

Hallegatte S (2013) An exploration of the link between development, economic growth, and natural risk. FEEM working paper 29. Milan, Italy

Hallegatte S, Green C, Nicholls R, Corfee-Morlot J (2013) Flood losses in major coastal cities. Nat Clim Chang 3(9):802–806

Hochrainer S (2009) Assessing macroeconomic impacts of natural disasters: are there any? Policy Research working paper 4968. World Bank, Washington, DC

IPCC (2012) Managing the risks of extreme events and disasters to advance climate change adaptation. A special report of Working Groups I and II of the Intergovernmental Panel on Climate Change [Field CB, Barros V, Stocker TF, Qin D, Dokken DJ, Ebi KL, Mastrandrea MD, Mach KJ, Plattner G-K, Allen SK, Tignor M, Midgley PM (eds)]. Cambridge University Press, Cambridge/New York, 582 pp

Jaramillo CRH (2009) Do natural disasters have long-term effects on growth? Universidad de los Andes/Mimeo, Bogotá

Mendelsohn R, Emanuel K, Chonabayashi S, Bakkensen L (2012) The impact of climate change on global tropical cyclone damage. Nat Clim Chang 2:205–209

Munich RE (2011) Topics geo natural catastrophes 2010. Munich Reinsurance Company, Munich

Noy I (2009) The macroeconomic consequences of disasters. J Dev Econ 88(2):221–231

Noy I, Nualsri A (2007) What do exogenous shocks tell us about growth theories? University of Hawaii working paper 07–28, Manoa, HI, USA

Peterson TC, Hoerling MP, Stott PA, Herring S (eds) (2013) Explaining extreme events of 2012 from a climate perspective. Bull Am Meteorol Soc 94(9):S1–S74

Raddatz C (2009) The wrath of god: macroeconomic costs of natural disasters, World Bank Policy Research working paper 5039. The World Bank, Washington, DC

Rose A (2004) Economic principles, issues, and research priorities in hazard loss estimation. In: Okuyama Y, Chang S (eds) Modeling spatial and economic impacts of disasters. Springer, Berlin, pp 14–36

Skidmore M, Toya H (2002) Do natural disasters promote long-run growth? Econ Inq 40:664–687

Skidmore M, Toya H (2007) Economic development and the impacts of natural disasters. Econ Lett 94:20–25

Strobl E (2011) The economic growth impact of hurricanes: evidence from U.S. coastal counties. Rev Econ Stat 93(2):575–589

World Bank (2009) World development report 2010. Development and climate change. World Bank, Washington, DC

World Bank (2010) Natural hazards unnatural disasters. The economics of effective prevention. World Bank report, Washington, DC

World Bank (2013) World development report 2014. Risk and opportunity: managing risk for development. World Bank, Washington, DC

# Chapter 2
# What Is a Disaster? An Economic Point of View

**Abstract** This chapter focuses on the economic consequences of a disaster and discusses the definition of the "economic cost" of a disaster. It stresses that a natural disaster is not a natural event, but the combination of a natural hazard (e.g., a hurricane) with a human system that is exposed to it and suffers from damages and perturbations. It reviews concepts such as direct and indirect losses, market and non-market losses, and consumption and output losses. The chapter also describes some of the most important mechanisms that determine the economic consequences of a disaster, such as the response of prices or the propagation of impacts through supply chains. It reviews the various tools that have been developed to measure and assess output losses from disasters, covering econometric analyses, input-output and computable general equilibrium models. It concludes with a definition of economic resilience and stresses the fact that reducing disaster welfare impacts can be done by reducing direct losses or by building resilience.

**Keywords** Natural disaster • Economic cost • Economic models • Disaster risk management

**A natural disaster is not a "natural" event**. Human and natural systems are affected by *natural hazards*, such as earthquakes, storms, hurricanes, intense precipitations and floods, droughts, landslides, heat waves, cold spells, and thunderstorms and lightning.

**If a hazard affects a human system – from one house to one region – and causes sufficiently larger negative consequences to this system, the event can then be labeled as a natural *disaster***. But a disaster occurs only when there is the conjunction of a natural event – the hazard – and a human system, leading to negative consequences. As such, what we call a natural disaster is thus above all a social and human event (World Bank 2010).

**From an economic perspective, a natural disaster can be defined as a natural event that causes a perturbation to the functioning of the economic system, with a significant negative impact on assets, production factors, output, employment, or consumption**. There are multiple formal definitions. The Center

---

This chapter is based on a working paper coauthored with Valentin Przyluski.



S. Hallegatte, *Natural Disasters and Climate Change*,
DOI 10.1007/978-3-319-08933-1_2

for Research on the Epidemiology of Disasters (CRED) at the Catholic University of Louvain – that maintains the EM-DAT database, see Box 2.1 – defines a disaster as a natural situation or event which overwhelms local capacity and/or necessitates a request for external assistance. For a disaster to be listed in the EM-DAT database, at least one of the following criteria should be met: (i) 10 or more people are reported killed; (ii) 100 people are reported affected; (iii) a state of emergency is declared; (iv) a call for international assistance is issued.

**Defining the economic cost of a disaster also poses different theoretical and practical challenges**. This chapter discusses these problems, and summarizes the most important mechanisms that determine the cost of disaster. It does so by first explaining why the direct economic cost, i.e. the value of what has been damaged or destroyed by the disaster, is not a sufficient indicator of disaster seriousness and why estimating indirect losses is crucial. Then, it describes the main indirect consequences of a disaster and of the following reconstruction phase, and discusses the methodologies to measure them.

## 2.1 Defining the Economic Cost of Extreme Events

After each large-scale disaster, media, insurance companies and international institutions publish numerous assessments of the "cost of the disaster." These various assessments are based on different methodologies and approaches, and they often reach quite different results. In the US, for instance, a systematic analysis by Downton and Pielke (2005) showed that loss estimates differ by a factor of 2 or more for half of the floods that cause less than $50 million in damages. These discrepancies are in part due to technical and practical problems, but also to the multi-dimensionality in disaster impacts and their large redistributive effects. Depending on what is included or not in disaster cost assessments, indeed, results can vary greatly. But the purpose of these assessments is rarely specified, even though different purposes correspond to different perimeters of analysis and different definitions of what a cost is.

This confusion translates into the multiplicity of words to characterize the cost of a disaster in published assessments: direct losses, asset losses, indirect losses, output losses, intangible losses, market and non-market losses, welfare losses, or some combination of those. It also makes it almost impossible to compare or aggregate published estimates that are based on so many different assumptions and methods.

### *2.1.1 Direct and Indirect Costs*

Many authors (e.g., Pelling et al. 2002; Lindell and Prater 2003; Cochrane 2004; Rose 2004) discuss typologies of disaster impacts. These typologies usually distinguish between direct and indirect losses.

**Direct losses are the immediate consequences of the disaster physical phenomenon: the consequence of high winds, of water inundation, or of ground shaking**. Typical examples include roofs that are destroyed by high winds, cars destroyed and roads washed away by floods, injuries and fatalities from collapsed buildings.

**Direct losses are often classified into direct market losses and direct non-market losses**.

**Market losses are losses related to goods and services that are traded on markets, and for which a price can easily be observed. For most disasters, direct market losses are losses of assets, i.e. damages to the built environment and manufactured goods**. They include the houses and buildings that are damaged or destroyed, the content of these buildings and houses (furniture, equipment, paper and data, etc.), and infrastructure (roads, bridge, etc.). These losses can be estimated as the repairing or replacement cost of the destroyed or damaged assets. Since building and manufactured goods can be bought on existing markets, their price is known: when a road is damaged, it is not difficult to estimate the cost of repairing it. Direct market losses can thus be estimated using observed prices and inventories of physical losses that can be observed or modeled (see Box 2.1 and an example in Table 2.1). Natural hazards also affect economic output, because offices and factories are closed during a storm for instance. For some hazards, such as heat waves and droughts, the main direct impact is on the economic output, not on assets: for instance a drought might not cause large damages to assets[1] but it can nevertheless reduce significantly agricultural production.

**Non-market direct losses include all damages that cannot be repaired or replaced through purchases on a market. For them, there is no easily observed price that can be used to estimate losses**. Non-market losses include health impacts and loss of lives, which are obviously a major component of natural disaster consequences. For instance, droughts can have permanent negative consequences on children development (e.g., diminished cognitive abilities) and floods are known to have large psychological impacts through post-disaster trauma. Disasters also damage historical and cultural assets, such as cathedrals and paintings, which have a high patrimonial value and are sometimes not exchanged on a market. Finally, disasters have impacts on natural assets and ecosystems, for instance when a hurricane leads to leaks of chemical products in the natural environment.

**It is difficult to attribute a cost to non-market impacts, since they cannot be "repaired" or "compensated" through financial transfers**. Sometime, a price for non-market impacts can be built using indirect methods, but these estimates are rarely consensual (more on this in Chap. 6).

One crucial aspect of disasters is that direct losses are not homogenously distributed. Investigating the 2004 hurricane season in Florida, McCarty and Smith

[1]Through its effect on soil dynamics, a drought may however cause large damages to buildings. The 2003 heat wave and drought over France is estimated to have cause damages to building larger than 1 billion euros.

(2005) find that – in their study area – 74 % of housing units were damaged, but only 2.2 % were totally destroyed while 40 % had only minor damages. Looking at the Northridge earthquake in 1994 in Los Angeles, Tierney (1997) finds that the median dollar loss from physical damage is US$5,000 while the average loss is US$156,000. Of course, this heterogeneity depends on the hazard type: losses from hurricane winds are more homogeneous than flood losses, which can vary dramatically depending on the topography. **These results show that disaster damages are heterogeneous, with many small losses and few large losses, making average and aggregated loss estimates poor indicators of welfare impacts**.

**Box 2.1: Available Data on the Economic Cost of Disasters**

The emergency Events Database (EM-DAT) maintained by the Center for Research on the Epidemiology of Disasters (CRED) at the Catholic University of Louvain, Belgium (http://www.emdat.be) is an important source of publicly available data on natural disasters. This database is compiled from diverse sources such as international financial agencies (e.g., the World Bank), UN agencies, NGOs, insurance companies, research institutions and press agencies.

The amount of damage reported in the database consists only of direct damages (e.g., damage to infrastructure, crops, and assets). The data report the number of people killed, the number of people affected, and the dollar amount of direct damages for each disaster.

Reinsurance companies also provide an extensive source of data but with important limits:

- The data are not publicly available, or only in an aggregated fashion;
- Reinsurance companies are collecting data based on the losses insured, and are thus biased toward countries where insurance is well developed (i.e. rich countries).
- These data usually disregard all indirect and nonmonetary losses.

Some loss data are also produced by "catastrophe models," developed to help insurers and reinsurers estimate natural disaster risks and set reinsurance premium. Such models exist where private reinsurance markets are well developed. They can estimate the losses caused by an event (e.g., on hurricane or one earthquake), using inventories of insured assets and models that predict the amount of damages caused by physical hazards (e.g., the value of damages to a house when wind speed exceeds 100 km/h). These models are usually run after each large-scale disaster to produce a first estimate of direct losses.

In the 1990s, the Economic Commission for Latin America and the Caribbean developed a formal methodology to assess disaster impacts, including indirect impacts. The ECLAC methodology (UN ECLAC 2003) assesses these impacts by a collection of data and information of various types

(continued)

**Box 2.1** (continued)
(physical, monetary, and expert judgment) in each sector (see an example in Table 2.1). Indirect costs are estimated through collection of information from economic agents, governments and experts, taking in consideration particular aspects such as transport disturbances cost, loss of opportunities, etc. Of course, data from different sources need to be aggregated carefully to avoid gaps and double counting.

**Indirect losses (also labeled "higher-order losses" in Rose** 2004**) include all losses that are not provoked by the disaster itself, but by its consequences; they are spanning over a longer period of time than the event, and they affect a larger spatial scale or different economic sectors.**[2] Like direct losses, indirect losses can be market or non-market losses. They include several categories of losses, such as:

- *Emergency costs*, i.e. the cost of intervention in the short term, which can range from a few hours for small events to months in case of large scale disasters like Katrina in New Orleans in 2005 or the Tohoku Pacific earthquake in March 2011. These costs include search and rescue costs, medical costs, when taking care of many injured victims at the same time is needed. They can also include security issues, even though evidence suggests that disasters trigger more collaboration and mutual assistance than unrest and looting (Solnit 2009). These costs can be significant: after the landfall of Katrina in New Orleans, emergency costs have been estimated at US$8 billion.
- *Business interruptions, supply-chain disruptions, and lost production due to capital damages* often represent a large share of indirect losses. The Iceland volcano eruption in 2010 interrupted air transport for a week, i.e. canceled air transport over the North Atlantic even in absence of any capital loss. A damaged factory after a hurricane cannot produce until it is rebuilt or repaired, leading to output losses. Output losses are also due to complex interactions between businesses, such as production bottlenecks when one element of a supply chain is affected and paralyze the entire production process.
- *Macro-economic feedbacks*, include the impact of reduced final demand because consumers and businesses suffer from a reduced income (e.g., due to loss of jobs), and the effect of lost tax revenue on public demand.
- *Demand surge*, i.e. the increase in repair costs after large disasters, because of the lack of workers and materials compared with the increase in demand due to reconstruction needs.

[2]Unsurprisingly, different hazards communities have different approaches for defining indirect costs. Contentious issues may emerge around the edge of these definitions across hazard communities.

**Table 2.1** Losses in the housing sector after the 2010 floods in Pakistan (US$ million)

| Province | Housing stock | | Other | | | | Total losses |
|---|---|---|---|---|---|---|---|
| | Completely destroyed | Partially destroyed | Equipment | Water and sanitation infrastructure | Debris removal costs | Temporary shelter costs | |
| **AJK** | 3.2 | 4.6 | 1.2 | 0.2 | 0.1 | 0.3 | **9.6** |
| **Balochistan** | 65 | 2.7 | 18.5 | 1 | 1.9 | 7.8 | **96.9** |
| **FATA** | 1.1 | 1.4 | 0.7 | 0.1 | 0.1 | 0.2 | **3.6** |
| **Gilgit-Baltistan** | 3.9 | 0 | 0.8 | 0 | 0.1 | 0.3 | **5.1** |
| **KPK** | 96.6 | 67.2 | 37.8 | 5.6 | 4 | 14.3 | **225.5** |
| **Punjab** | 125.7 | 97.9 | 54.7 | 18.4 | 5.6 | 20.1 | **322.4** |
| **Sindh** | 499.8 | 111.3 | 196.2 | 28.1 | 19.9 | 70.1 | **925.4** |
| **Total** | **795.3** | **285.1** | **309.9** | **53.4** | **31.7** | **113.1** | **1,588.5** |

Source: Preliminary damages and needs assessment. Global facility for disaster reduction and recovery (www.gfdrr.org)

- *Long-term adverse consequences on economic growth* are also possible because of changes in risk perception (including over-reactions) that can drive investors and entrepreneurs out of the affected area.
- *Long term consequences of health effects, psychological trauma and social network disruption* represent an additional source of indirect losses, which are subject of a growing interest. Beyond the direct welfare loss, indeed, these effects can reduce individual productivity and slow down development, economic growth, and poverty reduction. The consequences of evacuation on well-being and social networks can be larger, especially for poor households. Some of them are particularly dependent on informal social networks (e.g., for childcare), and these networks can be destroyed if evacuation is not organized to maintain them (McCarthy et al. 2006).
- *The impact on poverty or inequalities* is also sometimes included in the indirect losses. The landfall of Katrina on New Orleans has renewed attention on the larger weather vulnerability of the poorest communities within a country, and on the inequality-widening effect of disasters (e.g. Atkins and Moy 2005; Tierney 2006). Rodriguez-Oreggia et al. (2009) show that municipalities affected by disasters in Mexico see an increase in poverty by 1.5–3.6 % point. Often, the poorest have little to lose in a disaster and the impact on their welfare is therefore invisible in aggregated economic statistics. The case study on floods in Mumbai in Chap. 6 illustrates this problem. If the aim of the assessment is to look at welfare impacts, focusing only on economic aggregates can be misleading.
- *The impact on security, cohesion, stability* is also important. The Katrina landfall highlighted long term local security aspects of disaster (e.g., on food security, individual security, civil unrest), with for instance a 70 % increase in crime rate between the pre- and post-Katrina periods (see Van Landingham 2007). Sometimes, disaster consequences on social cohesion and political stability can be large. Even though disasters are never the unique cause of political unrest or even violent conflict, they have been important triggers in some cases (e.g., Bangladesh in 1970 and 1971 after hurricane Bohla).

Some of these impacts can be captured using classical economic indicators, such as GDP. There are however several issues when using GDP change as an indicator for indirect losses. A first question deals with the spatial scale: for large countries, the scale of the event and the scale of GDP measurement are very different, and a large shock for local populations can hardly be visible on national GDP. It does not mean, however, that welfare impacts are negligible. Second, the capacity of GDP to be a good proxy for welfare is also discussed, see Box 2.2.

**Box 2.2: How to Measure Welfare? Moving Beyond GDP**

The limits of GDP as an indicator of economic performance are well known, and have been summarized in several recent reports (e.g., Stiglitz et al. 2010; OECD 2009). In particular, the "Commission on the Measurement

(continued)

**Box 2.2** (continued)

of Economic Performance and Social Progress" report recalls the main shortcomings of current GDP measures: (1) the difficulty in measuring quality improvement in goods and services, which may lead to under or overestimation of real income growth; (2) government-produced goods and services are measured through their input value only, which may lead to underestimation of output change if government productivity increases. But beyond these limits, the report makes the case for shifting from measuring economic production to measuring welfare. To do so, it recommends several changes, and four of them are particularly important to investigate natural disaster consequences:

- Focus on income and consumption instead of GDP: welfare depends more on income and consumption than on GDP. In particular, an increase in depreciation (e.g., because new capital goods like computers depreciate more rapidly than older ones) may translate into an increase in investment, and thus into an unchanged GDP and a reduced consumption. Everything else being unchanged, such a change is negative for welfare but is not recorded in GDP. The Net Domestic Product (NDP) is the GDP net of depreciation and therefore takes this effect into account. Economic growth measured by NDP can be significantly different from economic growth measured with GDP. Since natural disasters affect capital depreciation and "force" reconstruction investments, the difference between GDP and consumption is critical: often a disaster can increase GDP through reconstruction investments, but this increase should not be considered welfare-enhancing.
- A specific case of depreciation is the depletion of natural resources. This depletion has direct consequences on our ability to use natural resources in economic production (e.g., water, oil). It has also a value linked to ethical consideration (our willingness to protect species and biodiversity, independently of the services they provide). Depletion could be captured by excluding the value of the natural resources harvested from the production value of sectors like mining and timber. Their production would then consist only in a pure extraction or logging activity, with a corresponding decrease in GDP. Or, depletion could be counted as depreciation, keeping GDP unchanged but reducing NDP. Disasters cause damages to natural resources and ecosystems, and taking into account natural capital in disaster assessments is thus very important.
- It is useful to take into account the concept of "defensive" expenditures (originally defined in Nordhaus and Tobin 1972). The cost associated to commuting is included in GDP, even though commuting does not yield direct welfare benefits but is only a requirement for other economic

(continued)

**Box 2.2** (continued)

activities. Many authors have proposed to treat these expenditures as intermediate consumption rather than final products, and to exclude them from GDP. But defining what is defensive is sometimes complicated. For instance, a new park can be considered as a defensive investment against the disamenities of urban life, or as a positive investment for recreational purposes. Because natural disasters force some defensive investments, for protection or reconstruction, the taking into account of defensive expenditures is crucial in the measure of their impact on welfare.

- Finally, many authors stress the importance of distribution in measuring economic growth and welfare, and suggest looking not only at averages but also at more sophisticated measures (from simple medians to quintiles). The same monetary loss would have very different consequences depending on whom it affects. Poor households will be particularly vulnerable to loss of assets, compared with wealthy households (Santos 2007; Alderman et al. 2006). This effect can be largely reinforced if impacts affect populations that are close to the subsistence level (Dercon 2004, 2005; Carter and Barrett 2006); in that case, even a monetary loss that is minuscule at the aggregate level can then have very large welfare impacts (see on Mumbai, Chap. 5). This aspect is important when investigating natural disasters: depending on the initial situation of the affected population and on how resulting costs are shared in a country or in a community, the total impact on welfare can be very different (see also a discussion of how to take this into account in a cost-benefit analysis in Chap. 6).

Many alternative indicators have been proposed – including the World Bank's Adjusted Net Savings that take into account natural resources and social capital – but there is no consensus on these indicators and GDP is still the most widely used for policy making and assessment.

An obvious illustration of why indirect losses are important is the difference between disaster scenarios with various reconstruction paces. In terms of welfare, there is a large difference between, on the one hand, a scenario in which all direct losses can be repaired in a few months thanks to an efficient reconstruction process and, on the other hand, a scenario in which reconstruction is inefficient and takes years.

### 2.1.2 *Defining a Baseline*

**A first difficulty in disaster cost assessment lies in the definition of the baseline scenario. The cost of the disaster has indeed to be calculated by comparing the actual trajectory (with disaster impacts) with a counterfactual baseline**

**trajectory (i.e., a scenario of what would have occurred in the absence of disaster)**. This baseline is not easy to define, and several baselines are often possible. Moreover, in cases where recovery and reconstruction do not lead to a return to the baseline scenario, there are permanent (positive or negative) disaster effects that are difficult to compare with a non-disaster scenario.

**For instance, a disaster can lead to a permanent extinction of vulnerable economic activities in a region, because these activities are already threatened and cannot recover, or because they can move to less risky locations**. In that case, the disaster is not a temporary event, but a permanent negative shock for a region. Also, reconstruction can be used to develop new economic sectors, with larger productivity, and lead to a final situation that can be considered more desirable than the baseline scenario. This improvement can be seen as a benefit of the disaster. It is however difficult to attribute unambiguously this benefit to the disaster, because the same economic shift would have been possible in the absence of disaster, making it possible to get the benefits without suffering from the disaster-related human and welfare losses.

For instance, hurricanes in 1806 and 1807 participated in the shift that occurred at la Réunion – a French island in the Indian Ocean – from coffee to sugar cane production. Indeed, it takes about 5 years for a coffee plant to start producing usable fruits. When the hurricanes hit, the need to start producing again as soon as possible drove farmers to plant sugar cane, an annual crop that can be harvested in a year. Moreover, coffee production was then considered more vulnerable to wind damages than sugar cane. But sugar cane production also has a different economic vulnerability to other stresses such as changes in food markets, consumer tastes, and agricultural subsidies. Therefore, it is difficult to assess whether the shift from coffee to sugar cane should be considered as a cost of the hurricanes (if sugar cane eventually reveals less profitable than coffee) or as a benefit (if sugar cane is more profitable than coffee). The final outcome depends on many other factors that can hardly be predicted at the time of the disaster.

This baseline issue – very common in economics – is not easy to deal with, and different scholars have used different techniques. Coffman and Noy (2012) use two nearby islands to assess the impact of hurricane Iniki. Since the hurricane affected only one island, the other can be used as a "control", i.e. as a proxy for the economic condition of the affected island if it had been spared. In the absence of a similar control, duPont and Noy (2012) use a statistical method to build a counter-factual for the Kobe economy, i.e. a scenario for the Kobe economy if no earthquake had occurred in 1995.

### 2.1.3 Assessment Purpose and Scope

**Defining the cost of a disaster cannot be done independently of the purpose of the assessment. Depending on the purpose of the assessment, some of the cost components have to be included or not in the analysis**. Different economic agents, indeed, are interested in different types of cost. Insurers, for instance, are mainly interested in consequences that can be insured. Practically, this encompasses mainly the cost of damages to insurable assets (e.g., damaged houses and factories), and short-term business interruption caused by the disaster (e.g., the impossibility to produce until electricity is restored).

For affected households, insurable assets are also a major component, but other cost categories are as important. Primarily, loss of lives, health impacts and perturbation to their daily life are crucial. In addition, households are concerned about their assets but also about their income, which can be reduced by business interruption or by loss of jobs, and about their ability to consume, i.e. the availability of goods and services.

At the society scale, all these aspects are important, but local authorities, governments and international institutions are also interested in other points. First, to manage the recovery and reconstruction period and to scale the necessary amount of international aid, they need information on the aggregated impact on economic production, on unemployment and jobs, on the impact of inequality and poverty, on local-businesses market-shares, on commercial balance, on collected taxes, etc. Second, to assess whether investments in prevention measures are desirable, they need the broadest possible assessments of the total disaster cost to the population, i.e. an estimate of welfare losses.

Moreover, disaster impacts can have positive or negative ripple-effects at the many scales. A neighboring region can benefit from the disaster because it captures market shares lost by the region hit by the disaster. The global economy can suffer from price effects, as shown by the significant rise in world oil prices that followed the landfall of hurricane Katrina in the Gulf Coast in 2005. Depending on the purpose and of the decision-making spatial scale, the perimeter of the cost analysis will be different. For instance, a country may want to assess the losses in the affected region, disregarding all out-of-the-region impacts, to calibrate the financial support it wants to provide to the victims. But a country may also want to assess total losses on its entire territory, including gains and losses outside the affected region, for example to assess the impact on its public finances.

## 2.2 Output Losses and Their Drivers

Damages to assets make them unable to produce: a damaged factory cannot build cars, a damaged road cannot be used, and a damaged house cannot be inhabited. The first step in an assessment of output losses is to estimate how much output is lost because of these direct asset losses.

### 2.2.1 *From Asset Losses to Output Losses*

**Economic theory states that, at the economic equilibrium and under certain conditions, the value of an asset is the net present value[3] of its expected future production. In this case, the annual loss of output is equal to the value of the lost capital multiplied by the interest rate (which is equal to the marginal productivity of capital). Assuming this equality is always verified, the output loss caused by capital loss is simply equal to the value of the lost asset, and summing the two is a double count.**

Figure 2.1 illustrates this point in a scenario in which no reconstruction takes place: in that case, the production that is lost because of the disaster is equal to the value of the lost assets. This equality between output losses and capital losses is however based on strong assumptions, which are not always verified.

In estimates of disaster consequences, *what is referred to as "asset loss" is the replacement value of the capital.* To have the equality of asset loss and output loss, a double equality needs to be verified:

- Replacement value has to be equal to market value; and
- Market value has to be equal to the net present value of expected output.

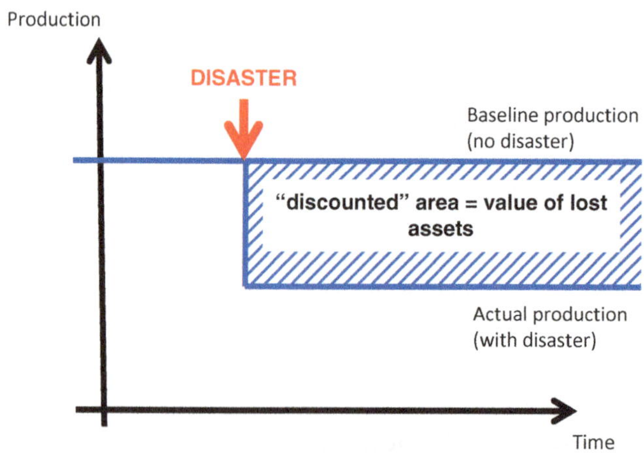


**Fig. 2.1** Production as a function of time, without disaster or in a scenario with disaster and no reconstruction. In the latter case, the discounted value of the lost production (from the disaster to the infinity) is equal to the value of lost assets. The production decrease is equal to the value of lost assets multiplied by the interest rate

[3]Note that the net present value is the sum of the production, discounted to account for the fact that production far in the future has less value than more immediate production; see a discussion on the discount rate in Chap. 6 and especially Box 6.2.

In a theoretical and optimal economy at equilibrium – these two equalities are valid. First, if the market value of an asset is lower (resp. larger) than the net value of its output, then investors will buy more (resp. sell) more of this asset to capture the difference in value, making asset price increase (resp. decrease).

Second, if market value were higher (resp. lower) than replacement value, then investors would increase (resp. decrease) the amount of physical capital to restore the equality between market and replacement value (assuming decreasing returns).

In a realistic setting, however, these two assumptions are not always verified. The reasons why asset values and output losses can differ are discussed in the following.

#### 2.2.1.1 The Economy Is Not at Its Optimum

**For the replacement value and the market value to be equal, the economy needs to be at its optimum, i.e. the amount of capital is such that its return is equal to the (unique) interest rate**. This is not always the case especially in sectors affected by disasters. In some sectors, expectations can be heavily biased (e.g., in the housing market) and markets distorted, leading to large differences between capital returns and interest rate. This is also unlikely for infrastructure and public assets. Since these assets are not exchanged on markets, they have no market prices. Moreover, they are not financed by decisions of private investors using financial returns, but by government decisions through a political process taking into account multiple criteria (e.g., land-use planning objectives). **Furthermore, output losses need to be estimated from a social point-of-view. The equality between market value (for the owner) and expected output (for the society) is valid only in absence of externalities**. Some assets that are destroyed by disasters may exhibit positive externality. It means that their value to the society is larger than the value of the owner's expected output. Public goods have this characteristic, among which most infrastructures, health services, education services.

An example is provided by the San Francisco Oakland Bay Bridge, which is essential to economic activity in San Francisco and had to be closed for 1 month after the Loma Prieta earthquake in 1989. Its replacement value has no reason to be equal to the loss in activity caused by the bridge closure, because the bridge production is not sold on a market, the bridge has no market value, and the social return on capital of the bridge is unlikely to exhibit decreasing returns and is likely to be much higher than the interest rate. Another example is the health care system in New Orleans. Beyond the immediate economic value of the service it provided, a functioning health care is necessary for a region to attract workers (what economists call a "positive externality"). After Katrina's landfall on the city, the absence of health care services made it more difficult to reconstruct, and the cost for the region was much larger than the economic direct value of this service.

#### 2.2.1.2 The Shock Is Large ("Non-marginal" in Economic Terms)

**The equality of asset value and output is valid only for marginal changes, i.e. for small shocks that do not affect the structure of the economy and the relative prices of different goods and services. The impact is different for large shocks.**

Most assets have "decreasing returns", i.e. their productivity decreases with the total amount of asset. For instance, if there are one million cars in a city, the loss of one car is a marginal shock, and the output of this car should be equal (at the optimum) to the production cost of a car. But in practice, some cars have a larger productivity than others: some cars are driven 1,000 km per year while others are driven 80,000 km per year; clearly the latter car is more productive than the former. In economic theory, the least productive car – i.e. the one that is driven the shortest distance per year – has an output equal to the production cost of cars. All the other ones have a higher productivity. As a result, the destruction of one car – assuming the least productive one is destroyed – leads to an output loss equal to the replacement cost of the car. But the destruction of many cars will affect cars with various productivity levels, and leads to an output loss that is larger than the replacement value of these cars.

**Moreover, the equality of asset value and output depends on the assumption that the destructions affect the least productive assets only**. In the previous example, it is assumed that if one car is destroyed, then it is the least productive (i.e. the one that is driven the shortest distance per year). Or equivalently, it is assumed that the owner of the destroyed car will instantaneously buy the least used car to its owner (which makes sense because the former makes a more efficient use of the car than the latter). Box 2.3 explains why this is unrealistic and why it leads to an underestimation of the output losses due to natural disasters.

As already stated, economic theory suggests that for marginal shocks, the annual loss in output should be equal to the value of lost assets multiplied by the interest rate (i.e. by the marginal productivity of capital). **In practice, it is more realistic to assess the loss of output as the value of lost assets multiplied by the average productivity of capital. Using classical production function and parameters (see Box 2.3), this approximation leads to output losses equal to approximately three times the value of lost assets.**

**Box 2.3: The Use of Classical Production Function Leads to Underestimating Output Losses**

There are several possible biases resulting from disaster modeling using classical production function (Hallegatte et al. 2007). Production functions are classically used in economics to relate the inputs and the outputs of a production process. Often, the production function takes as inputs the amount of labor used in the production process (referred to as L) and the amount of capital (i.e., the value of all equipment used in the process, referred to as K), and gives the value of the production (expressed as Y):

(continued)

**Box 2.3** (continued)

$$Y = f(L, K).$$

Disasters mainly destroy the stock of productive capital and a natural modeling option to represent their consequences is to consider that they reduce instantaneously the total productive capital ($K_0 \rightarrow K_0 - \Delta K$). Figure 2.2 illustrates several ways of assessing the impact on production. The figure represents the production Y as a function of capital K. The production function is the blue line linking the origin of the graph to the point A. It is assumed that the pre-disaster situation is the point A, with capital $K_0$ and production $Y_0$.

The impact on production can be estimated using the marginal productivity of capital, i.e. the interest rate at the optimum. This case is shown in Fig. 2.2 as the point B. Point B is estimated using the orange line, which is the tangent to the production function at point A; its slope is the marginal productivity of capital (i.e., how much more production do I get if I increase capital by one unit). The production $Y_1$ is given by the orange line at the X-coordinate $K_0 - \Delta K$, and is the estimated residual production if the output loss is estimated by multiplying the value of the lost capital $\Delta K$ by the interest rate. It is also what is done when the net present value of all output losses is assumed to be equal to the value of lost assets.

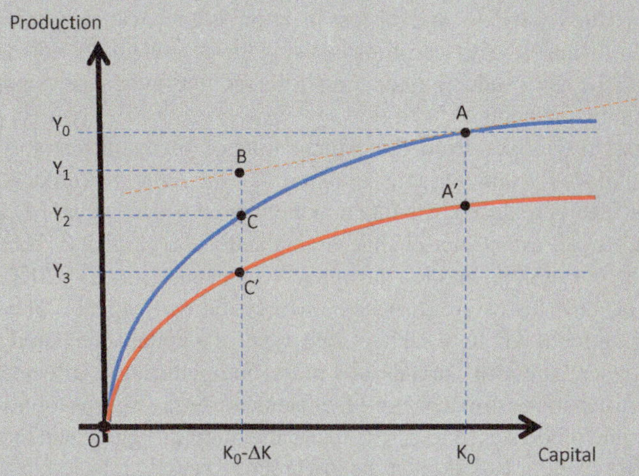


Fig. 2.2 Production with respect to productive capital for different modeling assumptions

(continued)

**Box 2.3** (continued)

The impact on production can also be estimated using the full production function, and decreasing the amount of capital from $K_0$ to $K_0 - \Delta K$. This is what is shown by the point C in the figure. The point C gives the value of production $Y_2$ given by the production function, i.e., $f(L,K_0 - \Delta K)$. This option, however, amounts to assuming that only the least-productive capital has been affected. Obviously, this is not the case: when a disaster hits, it destroys the capital indiscriminately, not only the least efficient capital.

Because of decreasing returns in the production function, using a classical production function amounts to assuming that capital destructions affect only the less efficient capital. In a Cobb–Douglas setting ($Y = AL^{\lambda}K^{\mu}$), indeed, the after-disaster production would be $Y_2 = AL^{\lambda}(K_0 - \Delta K)^{\mu}$. Since $\mu$ is classically estimated around 0.3, anx% loss of equipment would reduce the production by a factor ($\mu$ x), i.e. approximately (0.3 x) % (see figure below).

To account for the fact that disasters affect the capital independently of its productivity, Hallegatte et al. (2007) propose to modify the Cobb–Douglas production function by introducing a term $\xi_K$, which is the proportion of non-destroyed capital. This new variable $\xi_K$ is such that the effective capital is $K = \xi_K \cdot K_0$, where $K_0$ is the potential productive capital, in absence of disaster. The new production function is

$$Y_3 = \xi_K f(L, K_0) = A\xi_K L^{\lambda} K_0^{\mu}.$$

With this new production function, an x% destruction of the productive capital reduces production by x%, and the loss in output is approximately equal to 3.3 times ($=1/\mu$) the loss of asset estimated using the normal production function. In Fig. 2.2, the new production function is the red line and the new production $Y_3$ is given by the point C′.

Another bias arises from the aggregation of many different types of capital within only one variable – capital K – in economic models. If the function f(L,K) is replaced by a function with two types of capital $f(L,K_1,K_2)$, the impact of disasters can change dramatically. In particular, because of decreasing returns in $K_1$ and $K_2$, the impact of a given loss $\Delta K = \Delta K_1 + \Delta K_2$ depends on how losses are distributed across the two capitals. The loss in output is larger if all losses affect one type of capital, compared with a scenario where the two capitals are more homogeneously affected. As a result, disaster loss estimates can be dependent on the aggregation level of the economic models used to assess them: the more disaggregated the model is, the more likely it is that one type of capital is heavily affected, leading to large output losses (see also Sect. 2.4.2).

The use of production functions may create another problem: production functions assume that the output of the production process is continuous in

(continued)

**Box 2.3** (continued)
K and L. In reality, there are discontinuities in the production function: the loss of a segment of a road can make the entire road impracticable and useless; damages to one small piece of equipment in a factory can make it unable to produce the final product, etc. So a small ΔK can lead to a large loss in output, if the complementary of different capital items is taken into account. This is especially true when large network infrastructure is concerned.

**Asset and output losses are often estimated assuming unchanged (pre-disaster) prices, which is valid only for marginal shocks**. One can assume that if a house is destroyed, the family who owns the house will have to rent another house at the pre-disaster price. In other terms, the value of the housing service provided by the house can be estimated by the rental cost of a similar house before the disaster. But this assumption is unrealistic if the disaster causes more than a small shock. In post-disaster situations, indeed, a significant fraction of houses may be destroyed, leading to changes in the relative price structure. In this case, the price of alternative housing can be much higher than the pre-disaster price, as a consequence of the disaster-related scarcity in the housing market. For large shocks, estimating the value of lost housing service should take into account the price change (even though, as shown in Box 2.4, it is challenging to do so). Compared with an assessment based on the pre-disaster prices, it can lead to a significant increase in the assessed disaster cost. The same reasoning is possible in all other sectors, including transportation, energy, water, health, etc.

**In extreme cases, there may be rationing, i.e. the price cannot clear the market and supply is not equal to demand (Bénassy** 1993**). This is because markets are not at equilibrium in disaster aftermath**. The « If I can pay it, I can get it » assumption is not valid in post-disaster situations (e.g., there is no available house for rent at any price, there is no qualified worker to repair a roof). In these situations, therefore, the welfare impact of lost production cannot be estimated as the product of lost produced quantity and pre-disaster prices. Providing an unbiased estimate requires an assessment of the disaster impact on prices and taking into account rationing.

**Box 2.4: Quantity and Prices in Disaster Aftermaths**
Figure 2.3 is a classical quantity-price plot, showing the long-term demand and supply curves for a goods or service aggregated at the macroeconomic level. The green line is the demand curve: it shows how the quantity demanded by consumers decreases when the price increases. The blue curve is the pre-disaster (long-term) supply curve: it shows how the quantity produced

(continued)

**Box 2.4** (continued)

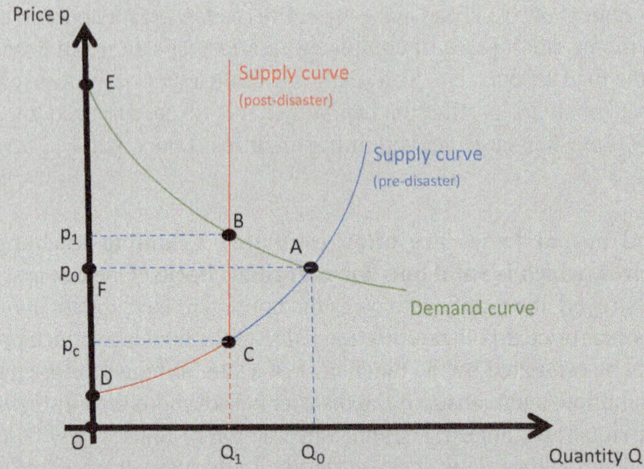


**Fig. 2.3** Supply and demand curves in the pre- and post-disaster situations

increases with the price (or, equivalently, the price asked by producers to produce a given quantity). The point A is the intersection of demand and supply and shows the price and quantity that clear the market (at that point, supply equals demand). The economic "surplus" is the area ADE. The consumer surplus is the upper area (AFE) and the producer surplus the bottom area (AFD).

The red line is the short-term supply curve after the disaster: because of damages, production cannot exceed $Q_1$, and the supply curve becomes vertical at this level (whatever the price consumers are ready to pay, producers cannot produce more than $Q_1$). If the market clears, the new equilibrium is reached at point B, where the quantity is reduced to $Q_1$ and the price increases to $p_1$.

In classical economic reasoning, the move from A to B is reducing the pre-disaster surplus ADE to the area BCDE. In other terms, the surplus loss is ABC. But this would be correct only if firms were deciding to reduce production from $Q_0$ to $Q_1$ *and to reduce the expenditure needed to produce* $Q_0$. If firms decided to reduce investment and production capacity from $Q_0$ to Q1, they would reduce their sales from $p_0Q_0$ to $p_1Q_1$, and reduce their production expenses from the area $ODAQ_0$ to the area $ODCQ_1$.

When a disaster hits, however, the sales are reduced from $p_0Q_0$ to $p_1Q_1$, but the expenses are not reduced from the area $ODAQ_0$to the area $ODCQ_1$. This is because firm expenses have three components: intermediate consumptions, capital expenses, and labor. The reduction in intermediate consumptions translates into a loss of output for another firm, so at the macroeconomic level,

(continued)

**Box 2.4** (continued)
a reduction in intermediate is not a gain (and intermediate consumptions are not included in the GDP). Reduction in labor expenditures is also a loss for workers, so it should not be counted at the macroeconomic level (unless, workers can find instantaneously another job, which is mostly not the case in disaster contexts). Finally, when a disaster reduces the production capacity from $Q_0$ to $Q_1$, it does not do so by reducing capital expenses, but by damaging existing capital. If a firm at a loan to pay for its capital (factory, equipment, etc.), the capital is destroyed but the loan is still there. In other terms, the capital expenditures are not reduced by the disaster.

So to assess the disaster impact on welfare over the short-term, it makes sense to consider the area $Q_0ABQ_1$ (and not the area ABC as in classical long-term welfare analysis). If the price is unchanged, then the impact can be estimated as $p_0\Delta Q$ (i.e. the loss of output). If the price change is significant, then it is necessary to take it into account, but it is challenging because the shape of the form $Q_0ABQ_1$ is complex. A linear assumption would simply be: $(p_0 + p_1)\ \Delta Q/2$.

**Post-disaster price inflation (also referred to as "demand surge") is especially sensible in the construction sector, which sees final demand soar after a disaster**. For instance, Fig. 2.4 shows the large increase in wages for roofers and carpenters in two areas heavily affected by hurricane losses in Florida in 2004. Demand surge is often considered as resulting from unethical behavior from businesses, justifying anti-gouging legislation (e.g., Rapp 2006). But demand surge can also have positive consequences. This inflation, indeed, helps attract qualified workers where they are most needed and creates an incentive for all workers to work longer hours, therefore compensating for damaged assets and accelerating reconstruction. It is likely, for instance, that higher prices after hurricane landfalls are useful to make roofers from neighboring unaffected regions move to the landfall region, therefore increasing the local production capacity and reducing the reconstruction duration. Demand surge, as a consequence, may also reduce the total economic cost of a disaster, even though it increases its financial burden on the affected population.

### *2.2.2 "Ripple Effects"*

Output losses are not only due to forgone production from the assets that have been destroyed or damaged by the event. Assets that have not been affected by the hazards can also reveal unable to produce at the pre-event level because of secondary effects, sometimes referred to as "ripple effects".

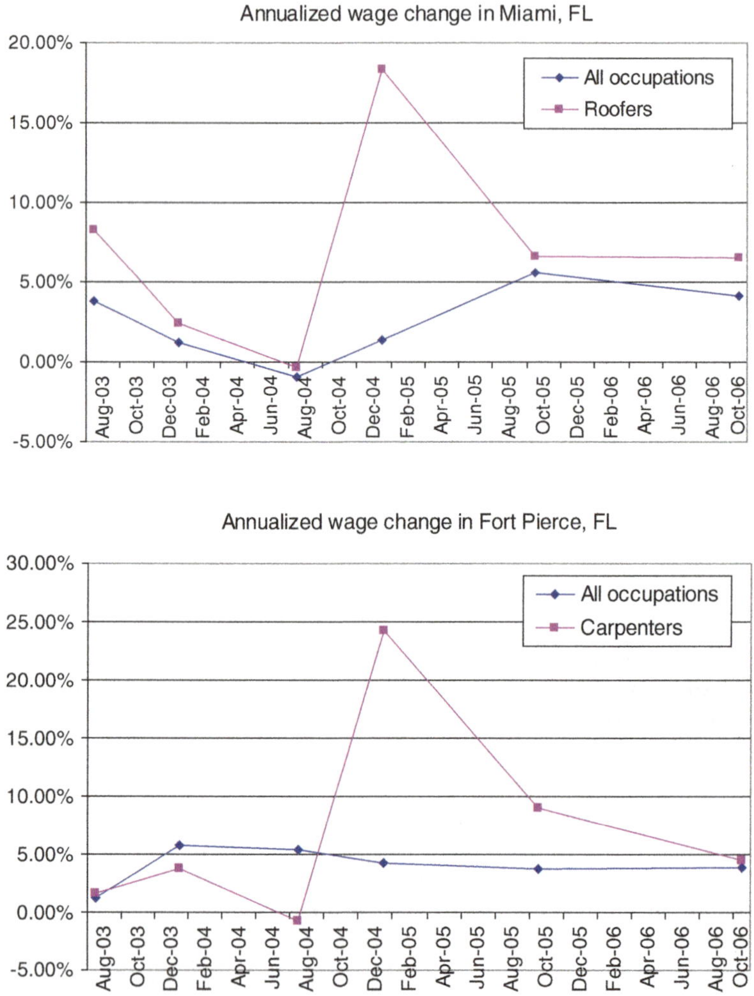


**Fig. 2.4** Wages for qualified workers involved in the reconstruction process (roofer and carpenter), in two areas where losses have been significant after the 2004 hurricane season in Florida (Data from the Bureau of Labor Statistics, Occupational Employment Surveys in May 03, Nov 03, May 04, Nov 04, May 05, May 06, May 07)

**This is particularly the case for infrastructure and utility services (electricity, water and sanitation, gas, etc.).** In past cases, it has been shown that the loss of utility services had larger consequences than direct asset losses, both on households (McCarty and Smith 2005) and on businesses (e.g., Tierney 1997; Gordon et al. 1998).

**McCarty and Smith** (2005) **investigated the impact of the 2004 hurricane season on households in Florida, and find that among the 21 % of the households who were forced to move after the disaster, 50 % had to do so**

**Table 2.2** Reason for businesses to close following the Northridge earthquake

| Reason | Percentage | Local (L) or indirect (I) |
|---|---|---|
| Needed to clean-up damage | 65.2 | L |
| Loss of electricity | 58.7 | I |
| Employees unable to get to work | 56.4 | I |
| Loss of telephones | 49.8 | I |
| Damage to owner or manager's home | 44.4 | I |
| Few or no customers | 39.9 | I |
| Building needed structural assessment | 31.5 | L |
| Could not deliver products or services | 24.0 | I |
| Loss of machinery or office equipment | 23.7 | L |
| Building needed repair | 23.4 | L |
| Loss of inventory or stock | 21.9 | L |
| Loss of water | 18.2 | I |
| Could not get supplies or materials | 14.9 | I |
| Building declared unsafe | 10.1 | L |
| Could not afford to pay employees | 9.5 | L |
| Loss of natural gas | 8.7 | I |
| Loss of sewer or waster water | 5.3 | I |
| Other | 15.8 | ? |

Source: Tierney (1997)

The reasons linked to local damages to the business are highlighted in *yellow*; the others are indirect reasons, due to perturbations in infrastructure services such as transport or electricity

**because of the loss of utilities (e.g., they had no running water). In only 37 % of the cases, the main reason was structural damages to the house**. Of course, households forced to move by direct losses to their home could not return as rapidly as the others, and they represent the majority of those who had to move out for more than a month (about 15 % of the households who had to move).

**Tierney** (1997) **and Gordon et al.** (1998) **investigate the impact of the Northridge earthquake in 1994 in Los Angeles; they find also that loss of utility services and transport played a key role**. Tierney surveys the reasons why small businesses had to close after the earthquake, see Table 2.2. The first reason, with 65 % of the answers (several answers were possible), is the need for clean-up. After that, the five most important reasons are loss of electricity, employee unable to get to work, loss of telephones, damages to owner's or manager's home, and few or no customers, with percentages ranging from 59 to 40 %. All these reasons are not related to structural damages to the business itself, but to offsite impacts. Gordon et al. (1998) ask businesses to assess the earthquake loss due to transportation perturbations, and find that this loss amounts to 39 % of total losses. Kroll et al. (1991) find comparable results for the Loma Prieta earthquake at San Francisco in 1989: the major problems for small business were customer access, employee access, and shipping delays, not structural damages. Utilities (electricity, communication, etc.) caused problems, but only over the short term, since these services have been restored rapidly; only transportation issues have led to long lasting consequences.

**Output losses are also due to complex interactions between businesses. Business perturbations may indeed also arise from production bottlenecks through supply-chains of suppliers and producers.**[4] These ripple-effects can be labeled "backward" or "forward:"

- **Backward ripple-effects** arise when the impact propagates from clients to suppliers, i.e. when a business cannot produce, and thus reduces its demand to its suppliers, reducing their own activity (even in absence of direct damages).
- **Forward ripple-effects** arise when the impact propagates from suppliers to clients, i.e. when a business cannot produce and thus cannot provide its clients with inputs needed for their own production (see Box 2.5).

**Some inputs are absolutely necessary for production, and a short interruption can cause significant perturbation in production.** Examples are electricity, water, fresh goods, and all other goods that are required for production and cannot be stocked. Other inputs are also necessary for production, but they can be stocked, and a short interruption in supply does not create large difficulties. An example is steel and tires for automakers, which are indispensable but stockable. Finally, other inputs are not absolutely necessary for production, and reasonably long interruptions can be managed. This is the case for instance of many business services, education and professional training.

**Inventories matter because they mitigate bottlenecks in the production system, introducing an additional – and critical – flexibility into the system.** Modern production organization tends to reduce the use of inventories with production-on-demand and just-in-time delivery. Added to other trends (e.g. outsourcing, reduction in the number of suppliers), these changes may make each production unit more dependent on the ability of its suppliers to produce in due time the required amount of intermediate goods. As a result, these changes may increase the overall vulnerability of the economic system to natural disasters, in a trade-off between robustness on the one hand, and efficiency in normal times on the other hand.

**One can show that the output losses due to a disaster depend on the characteristics of the firm-to-firm networks** (Henriet et al. 2012), such as the average number of suppliers that firms have, or the shape and structure of the connection between firms:

- *High-concentration* economies (i.e. economies where firms rely on a few suppliers only) are more vulnerable to disasters. This effect is magnified if firms have small inventories.
- *High-clustering* economies (i.e. an economy in which suppliers of a firm are likely to also be its clients) are less vulnerable to disasters; examples of clustered economies are "localized" economy, in which clients and suppliers are close to each other, compared to a "specialized–globalized" economy in which there are global supply chains.

---

[4]These ripple effects can even take place within a factory, if one segment of the production process is impossible and therefore interrupts the entire production.

These results suggest that modern economies, with global supply chains, limited number of suppliers and small stocks, may be more vulnerable to natural disasters than traditional, close economies. But the model used in Henriet et al. (2012) is too simple for providing realistic assessment of disaster costs, and detailed information on real-world economic networks is not available.

When capital cannot produce because of a lack of input (e.g., electricity, water), several options are available: input substitution, production rescheduling, and longer work hours can compensate for a significant fraction of the losses (see Rose et al. 2007). These mechanisms can damp output losses, and can especially reduce the crowding-out effects of reconstruction on normal consumption and investment. But their ability to do so is limited, especially when losses are large.

**Box 2.5: The Tohoku-Pacific Earthquake in Japan and the Auto Industry**

The impact of disasters on supply chains are tragically illustrated by the recent Tohoku-Pacific earthquake in Japan, and its wide consequences on industrial production and exports, especially in the auto industry. As an example, The Economic Times, an Indian newspaper, reports that "*Japan's Toyota Motor will cut production at its Indian subsidiary by up to 70 % between April 25 and June 4 due to disruption of supplies.*"

The New York Times also reports on the impact on Honda production: "*But auto production in Japan is at only half the normal level for Honda – as it is for Honda's bigger rivals, Toyota and Nissan. That is mainly because many of the 20,000 to 30,000 parts that go into a Japanese car come from the earthquake-stricken region in north-eastern Japan, where numerous suppliers were knocked off line. Unless part makers can resume production soon, the auto companies might have to shut down once again. 'We cannot continue for a long time,' said Ko Katayama, the general manager at Honda's factory here, declining to specify how long production could continue. Sooner or later, it's going to run out.*" These examples refer to highly-visible global supply chains, but such bottlenecks can also occur at the local scale, and even for small businesses.

**In case of large disasters, output losses will be largely dependent on two characteristics of the economy: the adaptability and flexibility of its production processes; and its ability to channel economic production toward its most efficient uses.**

**The "adaptability" and "flexibility" of the production system and its ability to compensate for unavailable inputs is largely unknown and largely depend on the considered timescale**. Over the very short term, the production system is largely fixed, and the lack of one input can make it impossible to produce. Moreover, over short timescales, local production capacity is likely to be highly constrained by existing capacities, equipment and infrastructure. Only imports from outside the affected region and postponement of some non-urgent tasks (e.g., maintenance) can create a limited flexibility over the short-term. Over the longer term and the entire reconstruction period, which can stretch over years for large-scale events, the flexibility is much higher: relative prices change, incentivizing production in scarce sectors; equipment and qualified workers move into the affected region, accelerating reconstruction and replacing lost capacities; and different technologies and production strategies can be implemented to cope with long-lasting scarcities. The production system organization can also be adjusted to the new situation: one supplier that cannot produce or cannot deliver its production (because of transportation issues, for instance) can be replaced by another supplier; new clients can be found to replace bankrupt ones; slightly different processes can be introduced to reduce the need for scarce inputs (e.g., oil-running backup generator can be installed if electricity availability remains problematic).

**Disaster indirect losses depend on the ability of the economic system to channel post-disaster residual production toward its best use, in order to minimize disaster losses**. Using existing production for reconstruction (instead to "normal" demand) accelerates reconstruction and reduces total losses (see Box 2.3). It is also more efficient to use output to satisfy intermediate demands from other sectors, to allow these other sectors to function and create value. The ability of the economic system to do so depends on how its institutions and markets can identify the "most important" demands (i.e. the demands that increase welfare most) and direct production toward these demands.

**First, the possibility to channel output toward the most productive use depends on the homogeneity of goods and services within each sector**. For instance, is it possible to divert some production that was directed toward final consumers to another industry intermediate demand? Doing so is possible if the produced goods are identical (e.g., the same car can be used by a final consumer or by another business); but in most cases, intermediate demands and final demands concern different goods and services (e.g., in cities, it is possible to have scarcity of residential housing but unused office space).

**Second, determining the optimal distribution of goods and services is indeed extremely difficult since it requires the taking into account of all supply-chains and business relationship in the economic system, i.e. a perfect knowledge of the economic network**. Theoretically, prices can play this role as they include all existing and necessary information. If prices were perfect, they would make it possible to optimize the use of remaining production capacities. For instance, if prioritizing intermediate consumptions over final demand increases total output and population well-being, then firms should be able to out-bid final consumers (i.e. they would be ready to pay a higher price) and they would capture the production they need. Over the short-term, however, prices are unlikely to include all information: they take a long time to adjust to new conditions, and different economic, governance and political processes have an influence on them. In particular, a producer that is unable to satisfy all the demand – and has to ration some of its (frustrated) clients – is very unlikely to also impose price hikes to them. Moreover, natural disasters are situations of abnormal solidarity and assistance (Solnit 2009) and price increases that reflect real scarcity will appear socially and politically unacceptable in disaster aftermaths.

### *2.2.3 Non-linearity in Output Losses and Poverty Traps*

**There are two main reasons why output losses are likely to increase non-linearly with the size of the disaster (and the amount of destruction). First, the "ripple effects" from infrastructure to firms and households and across firms are also likely to increase with the number of affected firms (and the individual loss of output).**

**Second, the reconstruction capacity is always limited by financial and technical constraint and it makes rebuilding after a large scale disaster much longer than after a small one**. In other terms, the duration of the shock increases with its amplitude. As a result, the output losses – that depend on the magnitude of the shock and its duration – will increase more than proportionally with direct losses.

The amount of damages can be a misleading indicator of the reconstruction duration. The 10 billion Euros of reconstruction expenditures after the 2002 floods in Germany correspond to 10 days of total German investments. But reconstruction has been spread out over more than 3 years, suggesting that only a small fraction of investments can be dedicated to reconstruction (even though the return on investment from reconstruction should theoretically be higher than other investments, as suggested by Box 2.3). One source of friction is that consumers, insurance and re-insurance companies, other companies and public organizations need time to direct high amounts of money to reconstruction activities. This constraint is crucial in developing economies (Benson and Clay 2004). Another source of friction is that the sectors involved in reconstruction activities have skills and organizational capacities adapted to the normal state of affairs and cannot face huge increases in demand (after the French storms in 1999 or after the AZF explosion in Toulouse in 2001, reconstruction took several years because roofers and glaziers were not numerous enough).

A model-based investigation of this issue using the Adaptive Regional Input-Output (ARIO) model (see a discussion on modeling in Sect. 2.4.2 and Hallegatte 2008, 2014) concludes that total losses due to a disaster affecting Louisiana increase nonlinearly with respect to direct losses when direct losses exceed \$50 billion (see Fig. 2.5). When direct losses are lower than \$50bn, aggregated indirect losses are close to zero (even though the aggregation hides important disparities among

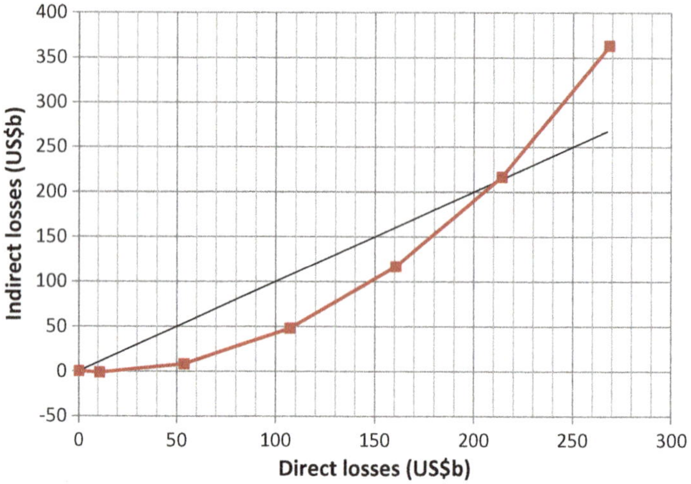


**Fig. 2.5** The direct losses – indirect (output) losses as a function of direct (asset) losses, in Louisiana for Katrina-like disasters of increasing magnitude

sectors and among social categories). Beyond $50 billion of direct losses, indirect losses increase nonlinearly. When direct losses exceed $200 billion, for instance, total losses are twice as large as direct losses. For risk management, therefore, direct losses are insufficient measures of disaster consequences.

**The output losses due to a hazard (and the resulting welfare impact) do not depend only on the physical intensity of the natural event, but also on the coping capacity of the affected human system. Physical measures of disaster intensity (e.g., in m/s for windstorm) or even measures of direct losses (e.g., the number and value of destroyed and damaged buildings) are very poor proxies of the real economic cost of a disaster.**

**A corollary is that natural disasters can have significant macroeconomic impacts, when they are larger than the (context specific) economic coping capacity. Moreover, it means that reducing natural disaster impacts on welfare can be done through an increase of the coping capacity, to make reconstruction faster and more efficient and limit output losses.**

This non-linearity could even lead to macro-level poverty traps, in which entire regions could be stuck. Such poverty traps could be explained by the amplifying feedback reproduced in Fig. 2.6: poor regions have a limited capacity to rebuild after disasters; if they are regularly affected by disasters, they do not have enough time to rebuild between two events, and they end up into a state of permanent reconstruction, with all resources devoted to repairs instead of addition of new infrastructure and equipment; this obstacle to capital accumulation and infrastructure development lead to a permanent disaster-related under-development.

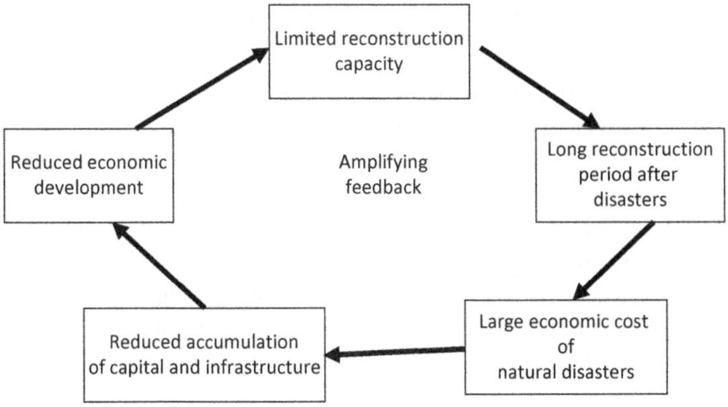


**Fig. 2.6** Amplifying feedback loop that illustrates how natural disasters could become responsible for macro-level poverty traps

This effect has been analyzed in Hallegatte et al. (2007)[5] that suggests that the long-term average GDP impact of natural hazards in one given region can be either close to zero if reconstruction capacity is large enough, or very large if reconstruction capacity is too limited (which may be the case in some least developed countries).

**In the model, there is a brutal "bifurcation" between two possible situations:**

- **Either disasters are small enough with respect to the reconstruction capacity, and disasters are short-term events with potentially large immediate impacts but a limited influence on long-term income and growth;**
- **Or disasters are too large (i.e. too frequent or too intense) with respect to the reconstruction capacity, and in that case disasters can have a significant impact on income and growth and development prospects.**

This type of feedback can be amplified by other long term mechanisms, like changes in risk perception that reduces investments in the affected regions or reduced services that make qualified workers leave the regions. Hallstrom and Smith (2005) assess the impact of hurricane risk perception on housing values in Florida, and find that hurricane risks reduces property values by 19 %. Such an impact on housing price can have broad consequence on economic activity, over the long term.

---

[5]This analysis was performed using the NEDyM model. This model is a highly idealized macroeconomic model that follows the classical Solow growth model in considering an economy with one representative producer, one representative consumer and one good, used both for consumption and investment. The original Solow (1956) model is composed of a static core describing the market equilibrium, and a dynamic relationship describing the productive capital evolution. In NEDyM, the static core is replaced by dynamic laws of evolution, building delays into the pathways toward equilibrium.

### *2.2.4 Building Back Better? The Productivity Effect*

**When a disaster occurs, it has been suggested that destructions can foster a more rapid turn-over of capital, which could yield positive outcomes through the more rapid embodiment of new technologies**. This effect, hereafter referred to as the "productivity effect" (also labeled "Schumpeterian creative destruction effect"), has been mentioned for instance by Albala-Bertrand (1993), Okuyama (2003), and Benson and Clay (2004).

Indeed, when a natural disaster damages productive capital (e.g., production plants, houses, bridges), the destroyed capital can be replaced using the most recent technologies, which have higher productivities.[6] Examples of such upgrading of capital are:

(a) For households, the reconstruction of houses with better insulation technologies and better heating systems, allowing for energy conservation and savings, and with a better resistance to natural disasters;
(b) For companies, the replacement of old production technologies by new ones, like the replacement of paper-based management files by computer-based systems, or new machines that are more effective and consume less energy;
(c) For government and public agencies, the adaptation of public infrastructure to new needs, like the reconstruction of larger or smaller schools when demographic evolutions justify it.

**Capital losses can, therefore, be compensated for by a higher productivity of the economy in the event aftermath, with associated welfare benefits that could compensate for the disaster direct consequences**. This process, if present, could increase the pace of technical change and accelerate economic growth, and could therefore represent a positive consequence of disasters.

**The productivity effect is probably not fully effective, for several reasons**. First, when a disaster occurs, producers have to restore their production as soon as possible. This is especially true for small businesses, which cannot afford long production interruptions (see Kroll et al. 1991; Tierney 1997), and in poor countries, in which people have no mean of subsistence while production is interrupted. Replacing the destroyed capital by the most recent type of capital implies in most cases to adapt company organization and worker training, which takes time. Producers have thus a strong incentive to replace the destroyed capital by the same capital, in order to restore production as quickly as possible, even at the price of a lower productivity. In extreme cases, one may even imagine that reconstruction is carried out with lower productivity, to make reconstruction as fast as possible, with a negative impact on total productivity (see how insurance can help in Box 2.6).

---

[6]On a related issue, it has been shown by Qing and Popp (2013) that natural disasters could trigger innovation in risk-mitigation technologies.

**Box 2.6: The Role of Insurance and Reinsurance**

The insurance industry plays a specific role in dealing with natural disasters. For small-scale events and uncorrelated risks, insurance is based on spreading risks among insurance customers. Because there is more or less the same number of car-insurance customers who have a car accident every year – the risk is said to be non-correlated –, it is possible to share this risk among them. Every year, the insurance premiums paid by all customers pay for the claims of those who have an accident. For individuals, the possibility of a large loss due to an accident is replaced by a certain and small payment every year. If agents are risk averse, this increases their welfare.

For correlated risks like large scale disasters, on the other hand, this principle no longer works: in any given year, there is either no disaster and there is no claim to pay, or one or more disasters and there are many claims to pay. In that case, the risk cannot be spread among customers of a single insurer portfolio in a single year, and need to be spread over time and over space. This is the role of reinsurance companies, which are global companies and keep a large amount of capital to pay a large number of claims in the same year in case of large-scale disaster(s).

The reinsurance companies are the insurers of the insurance companies. Their main activities include the coverage of frequency risk, catastrophe risk and longevity risks (in the case of life-insurance). Contracts, named treaties, are passed between a reinsurance company and the ceding insurer, determining the portfolio covered, the risk considered and the level of coverage. In the case of catastrophe insurance, the role of reinsurance is triple (1) it provides insurance companies with additional capacity to cope with correlated risks; (2) it allows for a worldwide mutualization of losses; (3) it provides an expertise on the risk insurers undertake. This third aspect of counseling particularly concerns insurance companies of medium and small sizes that do not have access to sufficient knowledge on the catastrophe risks.

Insurance is useful for several reasons:

- First, it spreads the risk among customers, over space, and over time. By doing so, it replaces rare and large losses by regular and small payments. If economic agents are risk averse, they prefer smooth and regular losses to an unknown and potentially large loss, and insurance increases welfare.
- If a disaster occurs, insurance claims support the affected population, reducing immediate welfare losses. Also, it helps fund the reconstruction and makes it shorter. As a consequence, output losses are reduced (see an illustration on Mumbai in Chap. 5).
- Because insurance provides a substitute income in disaster aftermath, it allows affected businesses and individuals to design and implement reconstruction strategies that take into account the most recent technologies.

(continued)

**Box 2.6** (continued)

Such reconstruction strategies are longer to undertake than recreating an identical production system. Affected agents, therefore, would carry them out only if they have an alternative source of income during reconstruction.

- Most economic agents are not aware of the risk they are facing. If insurance premiums are risk-based (which is not the case in most of the world), they provide an accessible measure of the level of risk. This signaling effect can help households and businesses make smarter choices (e.g., settling in lower-risk areas), and be an incentive to risk-mitigation actions (e.g., by reducing insurance premium if homeowners invest in flood-mitigation). In a situation of climate change, in which risk levels will change over time, a risk-based insurance premium may help disseminate information on changing climate conditions and be part of an adaptation strategy. The signaling effect of risk-based insurance premium should not be overstated, however. In many places, it would be too small to trigger significant behavioral changes.

Finally, any analysis of insurance should take into account affordability and solidarity issues. If current insurance premiums were transformed overnight into risk-based insurance premiums in Florida, many homeowners would face affordability issues. This is why some have proposed to combine the introduction of risk-based premium with the distribution of "insurance vouchers" to poor households, in an effort to mitigate adverse distributional impacts for the most vulnerable (Michel-Kerjan and Kunreuther 2011).

Also, it is important to note that there are limits to the amount of risk private actors can accept, and the highest risk levels (e.g., a large-scale earthquake in Japan, a hurricane on New York) will always remain with the government and public authorities. This risk transfer to public authorities can be made explicitly (like in France with the Cat-Nat system or in the US with the National Flood Insurance Program (NFIP)) or implicitly (like where the federal government helped non-insurers homeowners in Louisiana after the landfall of Katrina in 2005).

Second, even when destructions are quite extensive, they are rarely complete. Some part of the capital can, in most cases, still be used, or repaired at lower costs than replacement cost. In such a situation, it is possible to save a part of the capital if the production system is reconstructed identical to what it was before the disaster. This technological "inheritance" acts as a major constraint to prevent a reconstruction based on the most recent technologies and needs, especially in the infrastructure sector. This inheritance effect may for instance explain why it is almost never decided to reconstruct elsewhere – in the safer area – when a city located in a very risky location is affected.

In a simple economic model with embodied technical change (Hallegatte and Dumas 2008), disasters are found to influence the production level (i.e. average production is lower in areas with many disasters) but cannot influence the economic growth rate.[7] Depending on how reconstruction is carried out (with more or less improvement in technologies and capital), moreover, accounting for the productivity effect can either decrease or increase disaster costs, but this effect is never able to turn disasters into positive events.

Moreover, productivity is not only a function of the technology and physical capital that is installed. Disasters can have long-lasting consequences on psychological health (Norris 2005), and on children development (from reducing in schooling and diminished cognitive abilities; see for instance Santos 2007; Alderman et al. 2006). Migrations can also play a big role. They constitute a response to extreme events and disasters, especially in developing countries (McLeman and Smit 2006; Landry et al. 2007; Dillon et al. 2011). But they have consequences on long term development, especially if high-skilled, high-productivity workers are more able to migrate than the average population (as suggested by the Katrina case; see Zissimopoulos and Karoly 2007). These mechanisms can lead to a long-term reduction in labor productivity, potentially compensating gains from more recent capital.

### *2.2.5 The Stimulus Effect of Disasters*

**Disasters lead to a reduction of production capacity, but also to an increase in the demand for the reconstruction sector and goods. Thus, the reconstruction acts in theory as a stimulus. However, as any stimulus, its consequences depend on the pre-existing economic situation, such as the phase of the business cycles. If the economy is in a phase of high growth, in which all resources are fully used, the net effect of a stimulus on the economy will be negative, for instance through diverted resources, production capacity scarcity, and accelerated inflation. If the pre-disaster economy is depressed, on the other hand, the stimulus effect can yield benefits to the economy by mobilizing idle capacities.**

This complex interplay between business cycles and natural disasters economics is analyzed in detail in Hallegatte and Ghil (2008). The paper analyses disaster vulnerability in different phases of the business cycle in a simple model. In this model, this business cycle originates from the instability of the profit–investment relationship, a relationship similar to the Keynesian "accelerator–multiplier." When final demand is growing, firms need additional capacity to satisfy it, and they invest in new factories and other capacities. Doing that, they increase final demand

[7] This is similar to the fact that a change in the saving ratio in a Solow growth model can influence the output level, not the growth rate.

(because they buy buildings and factory and business equipment) and employment (because they hire workers). As a result, the initial increase in demand is amplified, leading to rapid growth. But part of this growth is unsustainable: it becomes more and more expensive to add capacity, with tensions on the labor market and decreasing returns of investments. At one point, profits become too low to justify additional capacity, and firms stop investing, leading to a drop in demand. At that point, the same mechanisms play, but in reverse: the decrease in demand due to lower investments means that there are excess capacities in the economy, and the economic stalls until wages have decreased and excess capacities have been resorbed (and profits are up again). The succession of these phases leads to the ups and downs of the business cycle.

**When affected by natural disasters, the model supports the counter-intuitive result that economies in recession are more resilient to the effects of natural disasters**. This result appears consistent with empirical evidence. For instance, the 1999 earthquake in Turkey caused direct destructions amounting to 1.5–3 % of Turkey's GDP, but consequences on growth remained limited, probably because the economy had significant unused resources at that time (the Turkish GDP contracted by 7 % in the year preceding the earthquake). In this case, therefore, the earthquake may have acted as a stimulus, and have increased economic activities in spite of its human consequences. In 1992, when hurricane Andrew made landfall on south Florida, the economy was depressed and only 50 % of the construction workers were employed (West and Lenze 1994). The reconstruction needs had a stimulus effects on the construction sectors, which would have been impossible in a better economic situation.

**There is a "vulnerability paradox":**

- **A disaster that occurs when the economy is depressed results in lower damages**, thanks to the stimulus effect of the reconstruction, which activates unused resources. In this situation, since employment is low, additional hiring for reconstruction purposes will not drive the wages up to a significant extent. Moreover, the inventories of goods during the recession are also larger than at equilibrium; a disruption of production, therefore, can be damped by drawing on them. In this case, the economic response damps the lasting costs of the disaster, according to the model.
- **A disaster occurring during the high-growth period results in larger damages**, as it worsens pre-existing disequilibria. First, the inventories are below their equilibrium value during the high-growth phase, so they cannot compensate for the reduced production. Second, employment is at a very high level and hiring more employees induces excessive wage inflation, thereby constraining economic activity.

This apparently paradoxical effect, however, is consistent with Keynes' ideas: "*If the Treasury were to fill old bottles with banknotes, bury them at suitable depths in disused coalmines which are then filled up to the surface with town rubbish, and leave it to private enterprise on well-tried principles of laissez faire to dig the notes up again [...] there need be no more unemployment [...]. It would indeed be more sensible to build houses and the like; but if there are political and practical difficulties in the way of this, the above would be better than nothing*." In fact, it is not much different to increase demand through fiscal or budgetary policies than through the destruction of houses and production capital that need to be replaced as rapidly as possible. In the latter situation, agents increase their consumption and reduce their savings, thus leading to a rise in demand. It is also well-known that this approach to macroeconomic stabilization is useless if supply is constrained, which is the case during the expansion phase of the business cycle.

**The stimulus benefits may be considered as a positive outcome of disasters, but the same effect would have been possible in the absence of a disaster, through a standard stimulus policy, and without the negative welfare and human impacts that come with disasters. So, it would be improper to attribute this positive effect to the disaster itself, and aggregating the corresponding benefits with disaster costs could be misleading on the real welfare cost of a disaster**.

## 2.3 From Output Losses to Welfare Losses

**Losses in economic output do not affect directly people welfare; for them, what matters most is consumption. It is thus important to investigate how output losses translate into consumption losses**.

Let's consider a scenario in which all losses are repaired instantaneously by reducing consumption and directing all the goods and services that are not consumed toward reconstruction investments (this is a scenario where reconstruction capacity is infinite). In this theoretical scenario, there is no output loss (since all asset damages are instantaneously repaired). There are however consumption losses, since consumption has to be reduced to reconstruct, and this reduction is equal to the reconstruction value (i.e. the replacement cost of damages capital). In that case, consumption losses are thus simply equal to the value of lost assets.

Let's now consider another scenario with no reconstruction, in which output losses are permanent (like in Fig. 2.1) and all losses in output are absorbed by a reduction in consumption. In that case, consumption losses are equal to output losses. In the theoretical case in which output losses are equal to the value of lost assets, therefore, consumption losses are identical in these two scenarios.

There is a continuum of scenarios between these two extremes, with reconstruction that can be very rapid or very long. But as long as output losses are equal to the value of lost assets, the loss in consumption is unchanged and equal to the value of lost assets.

**In the simplistic case where output losses are equal to the value of lost assets, the consumption losses are independent on the reconstruction scenario, and equal to the value of lost assets**. Section 2.2 showed however that output losses are larger than the value of lost assets, and that the instantaneous loss of output is larger than what is suggested by the marginal productivity of capital. Because of these differences, the two extreme scenarios (instantaneous reconstruction and no reconstruction) are not equivalent.

**In the instantaneous reconstruction scenario, consumption losses are equal to the share of consumption needed to repair and rebuild, i.e. the value of lost assets. In the no-reconstruction scenario, consumption losses are equal to output losses, i.e. larger than the value of lost assets.**[8] **As a result, consumption (and welfare) losses are magnified when reconstruction is delayed. The reconstruction phase, and the economic recovery pace, will ultimately determine the final cost of the natural disasters**. The fact that rapid reconstruction is better for welfare than slow reconstruction – or equivalently, that reconstruction has a return that is much higher than that of "normal investments" and the interest rate – explains why reconstruction is usually a priority and crowds out consumption and other investments in the affected region.

**A corollary is that it is possible to reduce the welfare impact of disasters by accelerating reconstruction. The ability to recover and reconstruct rapidly is often referred to as "resilience", and building resilience is one way of reducing the impact of disasters on welfare. Chapter 3 introduces a definition of risk that includes resilience, and Chap. 6 discusses recommendations for risk management "policy mix" that include actions to reduce direct and indirect losses.**

**As mentioned in Box 2.2, aggregated numbers can hide very large distributive impacts: GDP and aggregated consumption can be misleading instruments to measure actual welfare losses, especially on the poorest**. This is particularly important for disasters, as the poorest are usually more exposed and less able to protect themselves and recover from shocks. As already mentioned, Rodriguez-Oreggia et al. (2009) show that municipalities affected by disasters in Mexico see an increase in poverty by 1.5–3.6 % point. Baez and Mason (2008) present evidence for El Salvador, where earthquakes in 2001 reduced income and consumption per capita of the most affected households by one third. Various studies have investigated the 1984/1985 drought in Burkina Faso and find consumption levels reduced by up to 19 % in certain regions (Kazianga and Udry 2006; Fafchamps et al. 1998). These numbers show that the impact on affected households can be much larger than aggregated figures. Chapter 5 shows in a case study on Mumbai how these effects can be investigated, and confirms that aggregated impact analyses are not very useful to assess them. Then Chap. 6 discusses how cost-benefit analyses can account for the impact on poverty, and Chap. 7 shows that alternative decision-making techniques are able to account for poverty impacts directly, through the use of multiple metrics for measuring risk and disaster losses.

[8]The reality is more complex that what has been described here because not all output losses are translated into consumption losses. In practice, the loss in output changes the terms of the inter-temporal investment-consumption trade-off and translates into ambiguous instantaneous changes in consumption and investment. But the main conclusions of the analysis are not affected by this complexity.

## 2.4 Assessing Disaster Losses

### 2.4.1 *Measuring Indirect Losses Using Econometric Analyses*

**Econometrics analyses have been used to measure output losses, understood as reduction in GDP following a disaster, but they reach contradictory conclusions.** Albala-Bertrand (1993) and Skidmore and Toya (2002) suggest that natural disasters have a positive influence on long-term economic growth, probably thanks to both the productivity effect (Sect. 2.2.4) and the stimulus effect of reconstruction (Sect. 2.2.5). Others, like Noy and Nualsri (2007), Noy (2009), Hochrainer (2009), Jaramillo (2009), Raddatz (2009), and Felbermayr and Gröschl (2013) suggest that the overall impact of disasters on growth is negative. At local scale, Strobl (2011) investigates the impact of hurricane landfall on county-level economic growth in the US and shows that a county that is stuck by at least one hurricane over a year sees its economic growth reduced on average by 0.79 percentage point (and increased by only 0.22 percentage point the following year). On Vietnam, Noy and Vu (2009) investigate the impact of disasters on economic growth at the province level, and find that lethal disasters decrease economic production while costly disasters increase short-term growth.

**The lack of consensus on the impact of disasters on GDP is likely to arise from different impacts from small and large disasters,** the latter having a negative impact on growth while the former enhance growth, and from different impacts from different types of disasters. For instance, Felbermayr and Gröschl (2013) find that disasters in the top decile in terms of magnitude lead on average to a 3 % reduction in GDP growth. The loss is only 1.5 % for disasters in the top 15 % percentile, and 0.8 % for disasters in the top 20 %. For smaller disasters, no impact can be detected. The type of disaster also matters: Loayza et al. (2012) find that droughts reduce GDP growth by 1.7 percentage point, while floods increase GDP by 0.5 % (possibly because floods enrich soils and increase agricultural productivity).

This explanation appears consistent with results from modeling work: small disasters can create a stimulus effect that increases GDP in spite of the destructions; large disasters implies larger losses that cannot be compensated by a stimulus effect (see Sect. 2.2.5). Also, different disasters have different consequences: a drought reduces output without damaging assets, i.e. without creating reconstruction demands that can stimulate the economy.

### 2.4.2 *Modeling Indirect Losses*

Many scholars have used economic models to estimate output losses. Many different models have been used, but the most common are Input-Output (IO) or Calculable General Equilibrium (CGE) models. In these models, the economy is described as an ensemble of economic sectors, which interact through intermediate consumptions.

These models however describe differently how these different sectors interact with each other, and how they react to shocks.

**Some models are based on the Input-Output (IO) linear assumption (Leontief** 1951)**, in which the production of one unit in one sector requires a fixed amount of inputs from other sectors, and in which prices do not play any role**. This literature looks at different types of hazards (floods, earthquakes, droughts, or even terrorist attacks). It includes Rose et al. (1997); Haimes and Jiang (2001); Bockarjova et al. (2004); Okuyama (2004); Okuyama et al. (2004); Santos and Haimes (2004); Hallegatte (2008, 2014).

**The IO approach is based on the idea that, over the short term, the production system is fixed and that substitution possibilities are inexistent**. The IO model is originally a demand-driven model, which can be used to determine how much needs to be produced in each sector to satisfy a given level of final demand (see, e.g., Oosterhaven 1988; Cochrane 2004). A strict implementation of the IO approach implies that the most affected sector would become the bottleneck for the entire economic system, and that the final economic output would be reduced proportionally to the reduction in the most affected sector output. For instance if the plastic production sector experiences a 20 % decrease in production because one factory is damaged, this production loss would propagate into the economic system leading to a 20 % reduction in the final output of all sectors. This is an overestimation of the loss (for instance because it does not account for the fact that some intermediate consumption can be stocked and are not absolutely necessary for other production processes). Moreover, using this approach makes the final result in terms of output loss dependent on the sectoral disaggregation (the economy looks more vulnerable if it is described with many small sectors than with a few large sectors; see also Box 2.4).

This problem arises from the fact that IO models do not include a modeling of how to distribute the production (or the production scarcity after a shock) from one sector to different demands and different sectors. For example, if the plastic production is reduced by 20 %, it does not mean that all users and clients will experience a 20 % reduction in plastic production availability. Maybe some sectors (e.g., the automobile sector) will have access to an unchanged supply of plastic products while other sectors (e.g., the toy production sector) will experience a larger drop in supply. This is why **IO models used to model disasters are adapted to include a modeling of how the production of each sector is distributed across various clients and demands** (Okuyama 2004; Okuyama et al. 2004; Santos and Haimes 2004; Hallegatte 2008, 2014). But in these adjustments, the main assumption of the IO model remains: producing one unit in one sector (e.g., one car) requires a fixed and constant amount of inputs from other sectors (e.g., a given amount of steel, four wheels, etc.).

**Other models are based on the Calculable General Equilibrium (CGE) framework, which assumes that changes in relative prices balance supply and demand in each sector** (e.g., Rose and Liao 2005; Rose et al. 2007). In this framework, there is no rationing in the economic system and markets are always perfectly balanced. A disaster-caused destruction of production capital in a sector translates into a reduction in the production of the corresponding commodity, and into an increase in its price. This increase in price leads in turn to a reduction in its consumption, restoring the equality between demand and the reduced production. Moreover, in CGE models based on Cobb-Douglas or Constant Elasticity of Substitution production function, the production technology is not fixed anymore and there is short-term input substitution: in the presence of scarcity in one input, production can be carried out using less of this input, and using more of other inputs. Because of socio-economic inertia, transaction costs and anti-gouging legislation, however, the adjustment through prices appears unlikely in disaster aftermath. In post-disaster situations, little changes in prices have been observed, except in the construction sector. But prices and elasticity in CGE can also be seen as an artificial way of modeling flexibility. In that case, prices in the model should be considered as proxies for scarcity, more than actual observable prices. The CGE-model flexibilities (the signaling effect of prices, reduction in demand, and substitution in the production process) smooth any exogenous shock and mitigate disaster consequences.

**Economic losses caused by a disaster are smaller in a CGE setting than in an IO setting**. It is often considered that IO models represent the short-term economic dynamics, in which production technologies are fixed and prices cannot adjust. CGE models, on the other hand, represent the long-term dynamics, in which flexibility in production processes and markets allow for an adjustment of the economic system. In reality, it is likely that IO models are pessimistic in their assessment of disaster output losses, because there is flexibility even over the short term (for instance, maintenance can be postponed; workers can do more hours to cope with the shock; production can be rescheduled, see Rose et al. 2007). It is also likely that CGE models are optimistic, even in the long run, because prices have stickiness and cannot adjust perfectly, and because substitution has technical limits that are not always adequately represented in production functions.

**Natural disasters and their reconstruction phases are medium-term events, spanning from the first hours of the shock to years of reconstruction after large scale events**. Some authors have looked for intermediate approaches to natural disaster modeling, trying to find a common ground between IO and CGE. Rose and Liao (2005) and Rose et al. (2007) use a CGE framework, but with a lower substitution elasticity to take into account the fact that substitution is more limited over the medium term than over the long term. Going in the other direction, other authors developed IO models that answer previous IO-model shortcomings by introducing explicit supply constraints and production flexibility (e.g., Okuyama 2004; Okuyama et al. 2004).

**Another difficulty of disaster cost assessment lies in the aggregation level of CGE and IO models, which represent economic sectors, i.e. thousands of businesses located in different places, as a unique producer**. In such a framework, all businesses from one sector are assumed to suffer from the same direct impacts from disasters; in other terms, impacts are homogeneously distributed among the businesses from each sector. Also, if one business cannot produce, it is assumed that its production can be replaced by output from any other business of the same sector. These two assumptions are clearly overoptimistic, as they overestimate substitution capacity in the economic system. Taking into account the multiplicity of producing units, their location, and explicit supply-chains with inventories would allow for a much more realistic representation of natural disaster consequences. Haimes and Jiang (2001), Anderson et al. (2007), Battiston et al. (2007), Weisbuch and Battiston (2005), Coluzzi et al. (2010), and Henriet et al. (2012) investigate this issue and propose modeling approaches to account for these effects. But these works remain theoretical and far from any operational use.

**Considering the sensitivity of model results to many parameters, and the limitation of existing tools, it is fair to admit that models are useful tools to explore the indirect consequences of disasters but cannot estimate the cost of a disaster**. Progress in this domain would be welcome and useful. To do so, much more research should be devoted to this underworked problem.

## 2.5 Conclusion and the Definition of *Resilience*

This chapter highlights the main difficulties in defining, measuring, and predicting the total cost of disasters. It focuses on indirect (or output) losses, considered as a major component of the total loss of welfare.

**The main conclusion is that it is impossible to define "the cost" of a disaster, as the relevant cost depends largely on the purpose of the assessment**. The best definition depends on whether the assessment is supposed to inform insurers, cost-benefit analyses of prevention measure, or international aid providers. A first lesson is that any disaster cost assessment should start by stating clearly the purpose of the assessment and the cost definition that is used. Following this recommendation would avoid misleading use of assessments, and improper comparison and aggregation of results. Depending on the purpose of the assessment, the relevant definition of the indirect cost is different, and the most adequate methodology may also change.

**In particular, aggregated assessments provide national- or regional-scale assessments that can hide significant impacts on some categories, and especially on the poorest**. Beyond aggregated figures, it can therefore be recommended to complement classical economic analyses with targeted studies on how the poor (and other categories particularly exposed to some risk, such as farmers for droughts) are impacted.

**Also, the analysis shows that the welfare cost of a disaster does not depend only on the physical characteristics of the event, or on its direct impacts only. Depending on the ability of the economy to cope, recover and reconstruct, the reconstruction will be more or less difficult, and its welfare effects smaller or larger. This ability, which can be referred to as the *resilience* of the economy to natural disasters, is an important parameter to estimate the overall vulnerability of a population.**

To reduce the negative impact on natural disasters on population welfare, the first approach is to reduce the direct impacts on the economic systems, using for instance better coastal protections and stricter building norms. But another approach consist of reducing indirect losses through an increase in the resilience of the socio-economic system, using for instance insurance scheme or government support to the affected population. **An optimal risk management strategy is very likely to include measures targeting direct impacts (disaster risk reduction actions) and measures targeting indirect impacts (resilience building actions).**

## References


Albala-Bertrand JM (1993) The political economy of large natural disasters with special reference to developing countries. Clarendon Press, Oxford

Alderman H, Hodditnott J, Kinsey B (2006) Long-term consequences of early childhood malnutrition. Oxf Econ Pap 58:450–474

Anderson CW, Santos JR, Haimes YY (2007) A risk-based input-output methodology or measuring the effects of the August 2003 Northeast black-out. Econ Syst Res 19(2):183–204

Atkins D, Moy E (2005) Left behind: the legacy of Hurricane Katrina. Br Med J 331:916–918

Baez J, Mason A (2008) Dealing with climate change: household risk management and adaptation in Latin America. Background paper for the 2008 flagship report on climate change in Latin America and the Caribbean. World Bank, Washington, DC

Battiston S, Delli Gatti D, Gallegati M, Greenwald B, Stiglitz J (2007) Credit chains and bankruptcy propagation in production networks. J Econ Dyn Control 31:2061–2084

Bénassy JP (1993) Nonclearing markets: microeconomic concepts and macroeconomic applications. J Econ Lit 31:732–761

Benson C, Clay E (2004) Understanding the economic and financial impact of natural disasters. The international bank for reconstruction and development. The World Bank, Washington, DC

Bockarjova M, Steenge AE, van der Veen A (2004) On direct estimation of initial damage in the case of a major catastrophe: derivation of the "basic equation". Disaster Prev Manag 13: 330–336. doi:10.1108/09653560410556555

Carter MR, Barrett CB (2006) The economics of poverty traps and persistent poverty: an asset based approach. J Dev Stud 42:178–199

Cochrane H (2004) Economic loss: myth and measurement. Disaster Prev Manag 13:290–296

Coffman M, Noy I (2012) Hurricane Iniki: measuring the long-term economic impact of a natural disaster using synthetic control. Environ Dev Econ 17(2):187–205

Coluzzi B, Ghil M, Hallegatte S, Weisbuch G (2010). Boolean delay equations on networks: an application to economic damage propagation. arXiv preprint arXiv:1003.0793

Dercon S (2004) Growth and shocks: evidence from rural Ethiopia. J Dev Econ 74:309–329

Dercon S (2005) Vulnerability: a micro perspective. Paper presented at the annual bank conference on development economics. Amsterdam, 23–24 May 2005

Dillon A, Mueller V, Salau S (2011) Migratory responses to agricultural risk in northern Nigeria. Am J Agric Econ 93(4):1048–1061

Downton MW, Pielke RA (2005) How accurate are disaster loss data? The case of US flood damage. Nat Hazards 35(2):211–228

duPont W, Noy I (2012) What happened to Kobe? University of Hawaii, Working paper 2012-04. Manoa, HI, USA

Fafchamps M, Udry C, Czukas K (1998) Drought and saving in West Africa: are livestock a buffer stock? J Dev Econ 55(2):273–305

Felbermayr GJ, Gröschl J (2013) Naturally negative: the growth effects of natural disasters. SSRN scholarly paper no. ID 2348054. Social Science Research Network, Rochester, NY, USA

Gordon P, Richardson H, Davis B (1998) Transport-related impacts of the Northridge earthquake. J Transp Stat 1:21–36

Haimes Y, Jiang P (2001) Leontief-based model of risk in complex interconnected infrastructures. J Infrastruct Syst 7:1–12

Hallegatte S (2008) An adaptive regional input-output model and its application to the assessment of the economic cost of Katrina. Risk Anal 28:779–799

Hallegatte S (2014) Modeling the role of inventories and heterogeneity in the assessment of the economic costs of natural disasters. Risk Anal 34(1):152–167

Hallegatte S, Dumas P (2008) Can natural disasters have positive consequences? Investigating the role of embodied technical change. Ecol Econ 68(3):777–786

Hallegatte S, Ghil M (2008) Natural disasters impacting a macroeconomic model with endogenous dynamics. Ecol Econ 68(1–2):582–592

Hallegatte S, Hourcade J-C, Dumas P (2007) Why economic dynamics matter in assessing climate change damages: illustration on extreme events. Ecol Econ 62(2):330–340

Hallstrom DG, Smith VK (2005) Market responses to hurricanes. J Environ Econ Manag 50(2005):541–561

Henriet F, Hallegatte S, Tabourier L (2012) Firm-network characteristics and economic robustness to natural disasters. J Econ Dyn Control 36(1):150–167

Hochrainer S (2009) Assessing macroeconomic impacts of natural disasters: are there any? Policy Research working paper 4968. World Bank, Washington, DC

Jaramillo CRH (2009) Do natural disasters have long-term effects on growth? Universidad de los Andes/Mimeo, Bogotá

Kazianga H, Udry C (2006) Consumption smoothing? Livestock, insurance, and drought in rural Burkina Faso. J Dev Econ 79(2):413–446

Kroll CA, Landis JD, Shen Q, Stryker S (1991) Economic impacts of the Loma Prieta earthquake: a focus on small business, Studies on the Loma Prieta earthquake. University of California, Transportation Center, Los Angeles

Landry CE, Bin O, Hindsley P, Whitehead JC, Wilson K (2007) Going home: evacuation-migration decisions of Hurricane Katrina survivors. Center for Natural Hazards Research working paper, online at http://www.ecu.edu/hazards/pdfs/working_papers/Landry_etal.pdf

Leontief W (1951) The structure of American economy, 1919–1939, 2nd edn. Oxford University Press, New York

Lindell M, Prater C (2003) Assessing community impacts of natural disasters. Nat Hazards Rev 4(4):176–185

Loayza NV, Olaberria E, Rigolini J, Christiaensen L (2012) Natural disasters and growth: going beyond the averages. World Dev 40(7):1317–1336

McCarthy K, Peterson DJ, Sastry N, Pollard M (2006) The repopulation of New Orleans after Hurricane Katrina. RAND Gulf States Policy Institute, Santa Monica

McCarty C, Smith SK (2005) Florida's 2004 hurricane season: local effects. Florida Focus, University of Florida. http://www.bebr.ufl.edu/system/files/FloridaFocus1_3_2005_0.pdf

McLeman R, Smit B (2006) Migration as an adaptation to climate change. Clim Chang 76:31–53

Michel-Kerjan E, Kunreuther H (2011) Redesigning flood insurance. Science 333(6041):408–409

Nordhaus WD, Tobin J (1972) "Is growth obsolete?" Fiftieth Anniversary Colloquium V, National Bureau of Economic Research. Columbia University Press, New York

Norris FH (2005) Range, magnitude, and duration of the effects of disasters on mental health: review update 2005. Dartmouth Medical School and National Center for PTSD. Hanover, NH, USA

Noy I (2009) The macroeconomic consequences of disasters. J Dev Econ 88(2):221–231

Noy I, Nualsri A (2007) What do exogenous shocks tell us about growth theories? University of Hawaii working paper 07-28. Manoa, HI, USA

Noy I, Vu T (2009) The economics of natural disasters in Vietnam. University of Hawaii working paper 09-03. Manoa, HI, USA

OECD (2009) Measuring and fostering the progress of societies, document prepared for OECD meeting of the Council at Ministerial Level, 27–28 May 2010. C/MIN(2010)13, 12 May 2010. OECD, Paris

Okuyama Y (2003) Economics of natural disasters: a critical review. Research paper 2003-12. Regional Research Institute, West Virginia University

Okuyama Y (2004) Modeling spatial economic impacts of an earthquake: input-output approaches. Disaster Prev Manag 13:297–306

Okuyama Y, Hewings G, Sonis M (2004) Measuring the economic impacts of disasters: interregional input-output analysis using the sequential interindustry model. In: Okuyama Y, Chang S (eds) Modeling spatial and economic impacts of disasters. Springer, Heidelberg

Oosterhaven J (1988) On the plausibility of the supply-driven input-output model. J Reg Sci 28(2):203–217

Pelling M, Özerdem A, Barakat S (2002) The macro-economic impact of disasters. Prog Dev Stud 2:283–305

Qing M, Popp D (2013) Necessity as the mother of invention: innovative responses to natural disasters. National Bureau of Economic Research w19223

Raddatz C (2009) The wrath of god: macroeconomic costs of natural disasters, World Bank Policy Research working paper 5039. The World Bank, Washington, DC

Rapp GC (2006) Gouging: terrorist attacks, hurricanes, and the legal and economic aspects of post-disaster price regulation. Ky Law J 94:535. Available at SSRN: http://ssrn.com/abstract=800745

Rodriguez-Oreggia E, de la Fuente A, de la Torre R, Moreno H, Rodriguez C (2009) The impact of natural disasters on human development and poverty at the municipal level in Mexico. Manuscript

Rose A (2004) Economic principles, issues, and research priorities in hazard loss estimation. In: Okuyama Y, Chang S (eds) Modeling spatial and economic impacts of disasters. Springer, Berlin, pp 14–36

Rose A, Liao S-Y (2005) Modeling regional economic resilience to disasters: a computable general equilibrium analysis of water service disruptions. J Reg Sci 45:75–112

Rose A, Benavides J, Chang SE, Szczesniak P, Lim D (1997) The regional economic impact of an earthquake: direct and indirect effects of electricity lifeline disruptions. J Reg Sci 37:437–458

Rose A, Oladosu G, Liao SY (2007) Business interruption impacts of a terrorist attack on the electric power system of Los Angeles: customer resilience to a total blackout. Risk Anal 27:513–531

Santos I (2007) Disentangling the effects of natural disasters on children: 2001 earthquakes in El Salvador. Doctoral dissertation, Kennedy School of Government, Harvard University

Santos JR, Haimes YY (2004) Modeling the demand reduction input-output (I-O) inoperability due to terrorism of interconnected infrastructures. Risk Anal 24:1437–1451

Skidmore M, Toya H (2002) Do natural disasters promote long-run growth? Econ Inq 40:664–687

Solnit R (2009) A paradise built in hell. Penguin Group, New York, 368 pp

Solow R (1956) A contribution to the theory of economic growth. Q J Econ 70:65–94

Stiglitz JE, Sen A, Fitoussi JP (2010) Report by the commission on the measurement of economic performance and social progress. Commission on the Measurement of Economic Performance and Social Progress. The Commission, Paris

Strobl E (2011) The economic growth impact of hurricanes: evidence from U.S. coastal counties. Rev Econ Stat 93(2):575–589

Tierney K (1997) Business impacts of the Northridge earthquake. J Cont Crisis Manag 5:87–97

Tierney K (2006) Social inequality, hazards, and disasters. In: Daniels RJ, Kettl DF, Kunreuther H, Gutmann A (eds) On risk and disaster: lessons from Hurricane Katrina. University of Pennsylvania Press, Philadelphia

United Nations Economic Commission for Latin America and the Caribbean (UN ECLAC) (2003) Handbook for estimating the socio-economic and environmental effects of disasters. United Nations Economic Commission for Latin America and the Caribbean, Santiago

Van Landingham MJ (2007) Murder rates in New Orleans, La, 2004–2006. Am J Public Health 97:1614–1616

Weisbuch G, Battiston S (2005) Production networks and failure avalanches. arXiv preprint physics/0507101

West CT, Lenze DG (1994) Modeling the regional impact of natural disasters and recovery: a general framework and an application to Hurricane Andrew. Int Reg Sci Rev 17:121–150

World Bank (2010) Natural hazards unnatural disasters. The economics of effective prevention. World Bank report. Washington DC, USA

Zissimopoulos J, Karoly LA (2007) Employment and self-employment in the wake of Hurricane Katrina RAND working paper series WR-525. RAND, Santa Monica

# Chapter 3
# Disaster Risks: Evidence and Theory

**Abstract** Risk can be defined as the product of three factors: (1) the hazard, which is the natural event, such as a storm or an earthquake; (2) the exposure, which is the population and assets potentially affected by hazards; and (3) the sensitivity, i.e. the human and economic losses if population and assets are affected by a hazard. The chapter proposes a brief overview of natural risks today and discusses current trends in disaster losses by introducing the notion of risk-taking (and the benefits from risk-taking). Taking into account the benefits of taking risks helps us to understand why many people and assets are located in at-risk areas, such as flood plains and coastal zones. It also helps introduce potential policy options to reduce risk. The chapter concludes that there is no reason natural disaster losses will stop increasing in the future unless specific risk-reducing actions are implemented. Development does not automatically reduce natural disaster risks, and natural hazards need to be taken into account in development plans.

**Keywords** Disaster risk management • Disaster losses • Development • Risk-taking

While the previous chapter focused on the consequences of a single event (a hurricane, a flood), this chapter looks at *risk*, i.e. the possibility that a disaster hits. It starts by introducing a definition of risk, and then discusses current patterns of risk, including its spatial distribution and temporal trends. Then, it looks at the reasons why people create risk by investing in places that can be affected by floods and hurricanes, and proposes a framework to understand the rationale behind the observed patterns of risk-taking.

## 3.1 Defining Risk

It is widely recognized that economic losses due to natural disasters have been increasing exponentially in the last decades. However, disaster losses are not explained by natural events only; they depend also on socio-economic factors. Most of the time, **disaster risk is defined as the average annual disaster losses (or, equivalently, as the expected disaster losses for a given year).**


S. Hallegatte, *Natural Disasters and Climate Change*,
DOI 10.1007/978-3-319-08933-1_3

Disaster risk can then be decomposed into three explanatory factors, namely the hazard, the exposure, and the sensitivity:

- **The hazard probability, i.e. the likelihood of natural events that can cause disaster losses. A hazard is a physical event (e.g., a windstorm, an earthquake, a flood) and is measured using physical metrics (e.g., the maximum wind speed, the earthquake magnitude, the water level), which is independent of any socio-economic characteristics.** Each hazard (e.g., a windstorm with a maximum wind exceeding 100 km/h) can be attributed a probability. **This probability is often expressed as a "return period": for instance, the wind speed that has a 1 % chance of occurring or being exceeded over a 1-year period is referred to as the 100-year wind speed, because the average period between two of such events is 100 years.** For example, Fig. 3.1 shows the tropical storm hazard in the US; this information is independent of the localization of population and assets.
- **The exposure, i.e. the population and the amount of assets that are localized where the hazard can occur** (e.g., the population and houses located in a flood plain). The exposure depends on where population and assets are localized in the world. Figure 3.2 shows the exposure to tropical cyclones in 1970, with a projection for 2030.

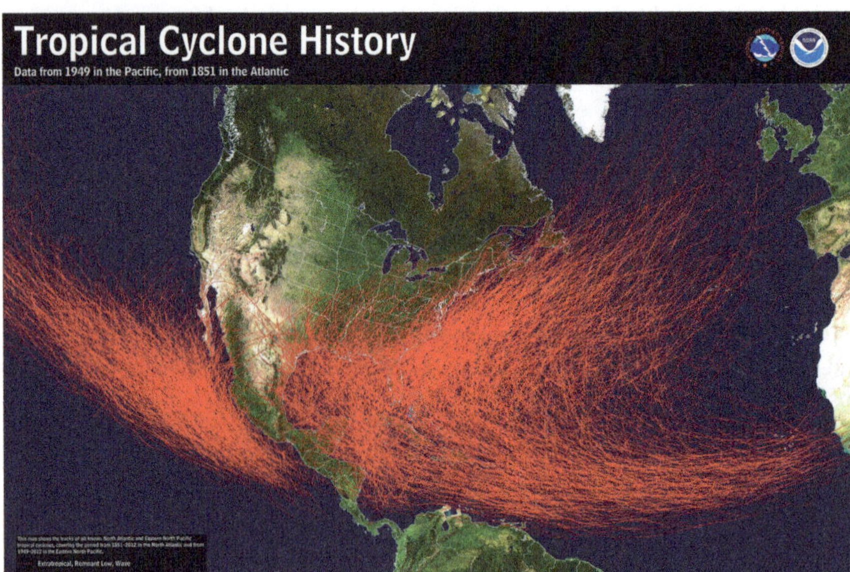


**Fig. 3.1** The tropical storm hazard in the US can be estimated using the density of past tracks (here from 1851 to 2013), according to NOAA

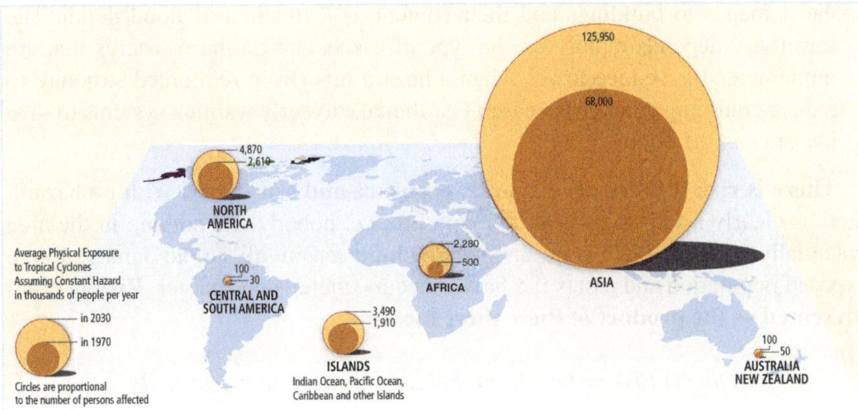


**Fig. 3.2** Average physical exposure to tropical cyclones assuming constant hazard (in thousands of people per year) (Data from Peduzzi et al. 2011. Figure from IPCC (2012); Fig. 4.1)

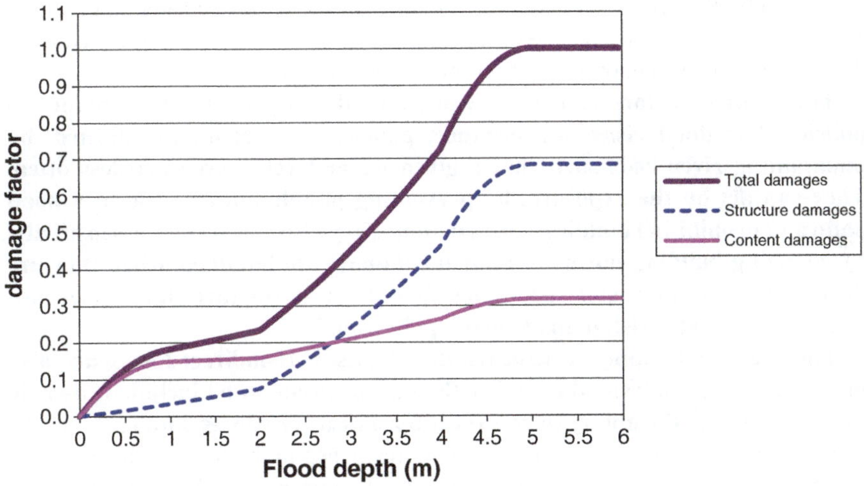


**Fig. 3.3** Relation between flood depth and damage factor for houses, distinguishing between damage to building and house content (Source: Jonkman et al. 2008)

– **The sensitivity (or vulnerability[1]), i.e. the amount of losses to the exposure, if a hazard occurs.** For instance, the sensitivity can be the amount of losses to a house – related to its full reconstruction value – if it is flooded or affected by high winds. Figure 3.3 provides an example for floods, with curves relating

[1]Vulnerability is defined differently in different communities. In the disaster risk management community, vulnerability is the sensitivity, i.e. the amount of damages in a house in case of flood. In the climate change community, the vulnerability is an integrated measure of the risk, including the probability of damage and the socio-economic capacity to respond.

the damages to buildings and their content as a function of flood depth. The sensitivity depends mainly on the type of assets and on the measures that are implemented to reduce losses when a hazard hits (from reinforced structure to reduced building damages in case of earthquake to early warning systems to save lives in case of flood).

**There is risk if there are hazards, exposure and sensitivity.** With no hazard, there is clearly no disaster. With no exposure, i.e. nobody and nothing in the area potentially affected, there is no disaster. With no sensitivity, i.e. no damage to the exposed population and assets if a hazard occurs, there is no disaster. **Risk is often presented as the product of these three factors:**

$$(Direct)\ Risk = hazard\ probability \times exposure \times sensitivity$$

For instance, the economic risk from the 100-year flood is equal to the probability of such an event (i.e., 1 % per year), multiplied by the economic value of the exposure (i.e., the economic value of the assets located in the 100-year flood plain), multiplied by the sensitivity (i.e., the ratio of losses to exposure value in case of a 100-year flood). In that case, the economic risk from the 100-year flood is equal to the average annual economic losses from 100-year floods.

**These three factors can also be used to design disaster risk reduction policies. For flood risks, for instance, policies can act on the hazard, by maintaining river beds such that a given water level is exceeded less often. They can act on the exposure, by preventing population to settle in a flood plain and prohibiting building construction. They also can act on the sensitivity, by changing building norms to limit the damages to buildings when they are flooded. Ideally, a disaster risk policy should act on the three levers (see also Chap. 6 on disaster risk management "policy mix").**

**These concepts apply to disaster direct losses. If indirect losses are also included in the analysis, then a fourth concept needs to be included, namely the resilience, or the ability to recover after a disaster causes damages.** Results from Chap. 2 show that taking into account resilience in the form of the ability to reconstruct is absolutely critical to assess output losses and welfare impacts. Some scholars have proposed to write the economic risk as the product of the hazard probability by the exposure and the sensitivity, divided by resilience:

$$(Welfare)\ Risk = hazard\ probability \times exposure \times sensitivity/resilience \qquad (3.1)$$

**In that case, risk is not defined based on the direct losses that can occur, but on potential welfare losses that can affect the population.** Using this framework, we can imagine two regions with the same hazard probability, exposure and sensitivity, but different resilience levels. For instance, region #2 may have introduced an efficient social safety net and post-disaster support systems to fasten recovery and reconstruction. In that case, the two regions have the same level of risk defined using direct losses only. But because recovery and reconstruction are more efficient in region #2, the welfare losses there might be only half of that in region #1. In that case, resilience in region #2 would be twice as high as in region #1.

It is relatively easy to assess the risk based on direct losses only (using inventory and localization of population and assets, and models to estimate their sensitivity to floods or earthquake). But as stressed in Chap. 2, assessing indirect losses is largely out of reach for current models, and resilience remains difficult to measure, assess, or predict. This is why this chapter focuses on direct risk, leaving aside for a moment the question of resilience (we will come back to it in Chap. 5).

## 3.2 The Current Patterns of Risk[2]

Natural catastrophes affect countries around the world, regardless of their level of economic development. Depending on geographical circumstances, developing and developed countries alike can be particularly prone to experience frequent natural "loss events"– for instance, Bangladesh (floods), the USA (hurricanes), or Japan (earthquakes). However, the nature of experienced damages and losses varies significantly subject to an individual country's level of income, and this is due to differences in terms of hazard probability (for instance, some countries are affected more often than others by hurricanes), exposure (for instance some countries have more population in flood zones), sensitivity (for instance, poor countries have housing of lower quality that leads to more casualties in case of earthquake), and resilience (for example, some countries are able to reconstruct more quickly than others in case of disasters).

**In analyzing the structure of disaster related losses, it becomes apparent that: (i) the vast majority of human fatalities are experienced by low-income countries; (ii) absolute economic losses are the highest in high-income countries; and (iii) economic losses relative to GDP are the highest in low-income countries, with further implications for e.g. the speed of recovery and reconstruction.**

Figure 3.4 shows the number of victims of natural disasters (per 100,000 inhabitants) in each country, over the 1976–2005 period. It shows that natural disasters affect primarily developing countries, and especially Asia, where several countries have more than 5 % of the population affected during the considered period. This is especially the case of India and China, two countries with very large populations.

[2]This section was contributed by Jun Rentschler.

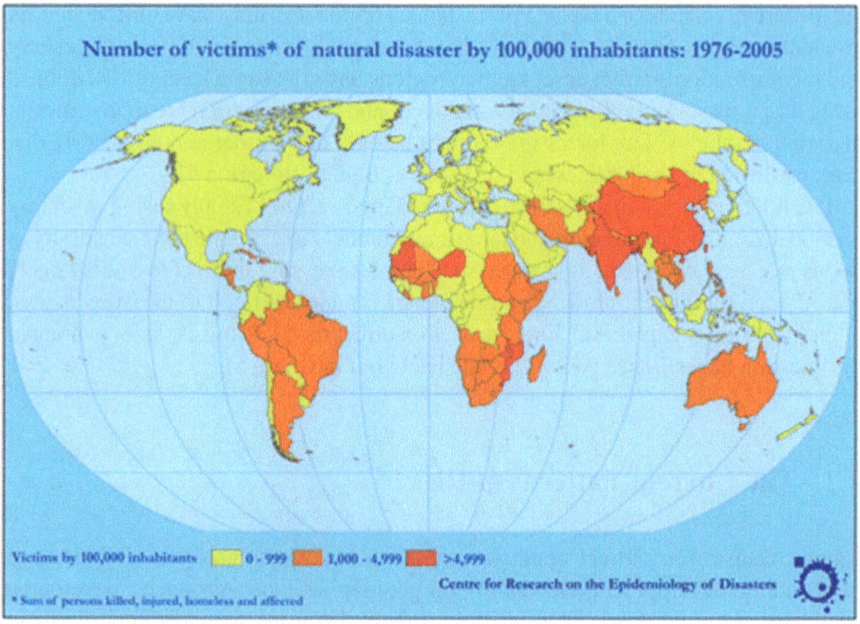


**Fig. 3.4** Number of victims of natural disasters per 100,000 inhabitants over the 1976–2005 period (Source: EM-DAT: The OFDA/CRED International Disaster Database – www.emdat.be – Université Catholique de Louvain – Brussels – Belgium)

In order to distinguish between different income levels, it is useful to apply the World Bank country classifications according to income groups[3]:

Income group 1 (IG1): High-income countries (Per capita income US$ > 12,275);
Income group 2 (IG2): Upper-middle-income countries (Per capita income US$ 3,976–12,275);
Income group 3 (IG3): Lower-middle-income countries (Per capita income US$ 1,006–3,976);
Income group 4 (IG4): Low-income countries (Per capita income US$ < 1,006).

Tables 3.1 and 3.2 present the most harmful natural catastrophes according to fatalities, as well as absolute economic losses between 1980 and 2012.

Considering these lists of the deadliest and the costliest natural catastrophes of the past 30 years, two features are particularly striking: The vast majority of human fatalities as a result of natural catastrophes are incurred in low-income, or lower-middle income countries (Table 3.1). On the contrary, the highest absolute economic losses from natural catastrophes are incurred in high-income economies (Table 3.2).

[3]For coherence with the subsequently presented data, we use the World Bank's 2011 country classification.

**Table 3.1** Disasters by fatalities (1980–2012)

| Year | Event | Location | Fatalities | Income group |
|---|---|---|---|---|
| 2010 | Earthquake | Haiti | 222,570 | IG4 |
| 2004 | Earthquake, tsunami | Sri Lanka, Indonesia, Thailand, India, Bangladesh, Myanmar, Maldives, Malaysia | 220,000 | IG4, IG3, IG2 |
| 2008 | Cyclone Nargis, storm surge | Myanmar | 140,000 | IG4 |
| 1991 | Tropical cyclone, storm surge | Bangladesh | 139,000 | IG4 |
| 2005 | Earthquake | Pakistan, India, Afghanistan | 88,000 | IG4, IG3 |
| 2008 | Earthquake | China | 84,000 | IG2 |
| 2003 | Heat wave, drought | France, Germany, Italy, Portugal, Romania, Spain, United Kingdom | 70,000 | IG2, IG1 |
| 2010 | Heat wave | Russia | 56,000 | IG1 |
| 1990 | Earthquake | Iran | 40,000 | IG2 |
| 2003 | Earthquake | Iran | 26,200 | IG2 |

Source: MunichRE, NatCatSERVICE

**Table 3.2** Disasters by absolute economic losses (in US$ m, original values, 1980–2012)

| Year | Event | Location | Losses (US$ m) | Income group |
|---|---|---|---|---|
| 2011 | Earthquake, tsunami | Japan | 210,000 | IG1 |
| 2005 | Hurricane Katrina, storm surge | USA | 125,000 | IG1 |
| 1995 | Earthquake | Japan | 100,000 | IG1 |
| 2008 | Earthquake | China | 85,000 | IG2 |
| 2012 | Hurricane Sandy, storm surge | USA, Canada, Bahamas, Cuba, Dominican Rep., Haiti, Jamaica, Puerto Rico | 65,000 | IG1[a] |
| 1994 | Earthquake | USA | 44,000 | IG1 |
| 2011 | Floods | Thailand | 43,000 | IG2 |
| 2008 | Hurricane Ike | USA, Cuba, Haiti, Dominican Rep., Turks and Caicos Islands, Bahamas | 38,000 | IG1[a] |
| 1998 | Floods | China | 30,700 | IG2 |
| 2010 | Earthquake, tsunami | Chile | 30,000 | IG1 |

Source: MunichRE, NatCatSERVICE

[a]The vast majority of economic losses were incurred in the USA, i.e., IG1

In fact, the deadliest natural catastrophe experienced between 1980 and 2012 by a high-income country (IG1) was the 2003 heat wave, which claimed 70,000 lives in Europe. This death toll however is much less than the deadliest catastrophe experienced by a low-income country (IG4) in the same time period: the 2010 earthquake in Haiti with 222,570 fatalities.

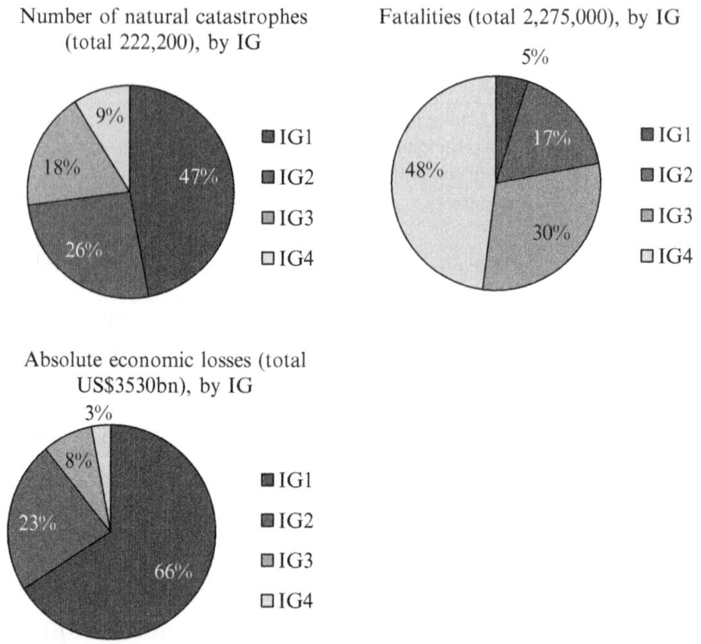


**Fig. 3.5** Distribution by income group (1980–2011); *Left*: Total number of loss events; *Middle*: Fatalities due to natural catastrophes; *Right*: Absolute economic losses in 2011 values (Source: MunichRE, NatCatSERVICE)

Considering all of the recorded natural catastrophes with human or economic losses between 1980 and 2011 offers a better understanding of how the harmful consequences are distributed across income groups. In this time period a total of 22,200 events were recorded, out of which nearly half were experienced in high-income countries (c.f. Fig. 3.5).[4] At 9 % low-income countries experienced a relatively small number of natural catastrophes in this time period. However, the distribution of fatalities offers a completely reversed image: nearly half of the 2,275,000 fatalities due to natural catastrophes in the same time period were incurred by countries in the lowest income group, while high-income countries incurred far fewer fatalities (5 %).

In terms of economic losses due to natural catastrophes, the picture also relies strongly on the applied measure and the considered income group: For the period from 1980 to 2011 absolute economic losses of US$3,530 billion were incurred globally, out of which two thirds were incurred by high-income countries (c.f. Fig. 3.5). Low-income and lower-middle income countries combined incurred just over 10 % of the total absolute losses (while incurring more than 75 % of total fatalities).

[4]Events are recorded at country level, i.e,. 2012 Hurricane Sandy is recorded as a separate event for each affected country.

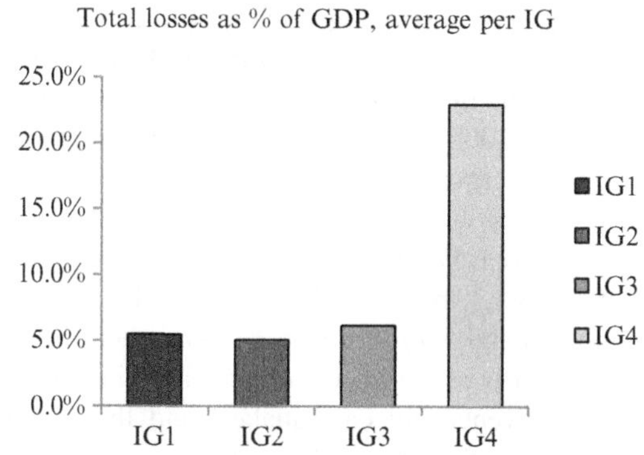


**Fig. 3.6** Economic losses as percentage of GDP, 1980–2011 (Source: MunichRE, NatCatSERVICE)

Relative to GDP however, low-income countries incurred on average the highest economic losses as a consequence of natural catastrophes (c.f. Fig. 3.6). Overall economic losses averaged close to one quarter of GDP in the lowest income group, while being between 5 and 8 % in the other income groups. This demonstrates that even though absolute economic losses were low in low-income countries, the economic burden of natural catastrophes is disproportionately higher in these countries, with further implication for e.g. the speed of economic recovery and means for reconstruction.

Skidmore and Toya (2007) use a cross-section analysis of 151 countries over the 1960–2003 period, with disaster loss data, and find that the ratio of economic losses to GDP is inversely related to GDP per capita. According to their analysis a 1 % gain in GDP per capita translates into a 0.5 % decrease in the loss to GDP ratio. If other drivers are accounted for (schooling years, size of government, openness, and financial development), this effect still leads to a 0.1 % decrease in loss to GDP ratio for each 1 % increase in GDP per capita. Bakkensen (2013) finds similar results, even accounting for country fixed-effects. This result confirms that rich countries suffer from most of the losses in absolute terms, while poorer countries have higher losses when compared with their resources. Figure 3.6 suggests moreover the existence of a threshold, as the ratio of losses to GDP remains relatively uniform for countries from lower-middle income to high income status, and is much higher for low-income countries. Schumacher and Strobl (2011) find that disaster losses can increase with wealth for low GDP per capita levels where the risk is low or moderate, and for all GDP per capita where the risk level is particularly high.

## 3.3 Current Trends

Considering the higher vulnerability of poor countries, one can expect that global economic growth and development will mitigate disaster risk in the future. But the trend in disaster losses is more complex and does not need to be monotonic.

**In spite of a growing world population and an increase number of disasters, the number of deaths from natural disasters is relatively stable, suggesting that populations are better protected than in the past** (Kahn 2005). Kellenberg and Mobarak (2008) find evidence that for three types of disasters – floods, landslides, and windstorms – disaster deaths increase with rising income before they decrease. In countries with a GDP per capita level below roughly $4500–$5500, disaster deaths increase in income. Beyond this wealth level, disaster deaths start decreasing with growing income. They explain this inverse-U-shaped curve by risk-related behaviors and varying levels of investment in disaster protection as income rises. This trend is also linked to better buildings that provide better protection against weather events and earthquakes, and the growing availability of early warning systems (see the case study on this topic in Chap. 6).

**For economic losses, there is a clear increasing trend in global disaster losses over time, and this trend is at least as fast as global economic growth.** With more and richer people, it is not surprising to find an increase in disaster losses. More surprising is the fact that, in spite of growing investments in risk reduction (i.e. a decrease in "sensitivity" in Eq. (3.1)), the growth in losses has been at par with economic growth. There is evidence for floods in Europe than losses have been growing at the same rate as GDP growth (Barredo 2009). At the global scale, with much larger uncertainties, Miller et al. (2008) or Neumayer and Barthel (2011) also find that losses grow as fast as GDP. In the U.S., Nordhaus (2006) and Pielke et al. (2008) find an increase in losses more rapid than GDP growth. This situation can be fully explained by the fact that people take more and more risks, by moving and investing more and more in at-risk areas, i.e. an increase in "exposure" in Eq. (3.1); see Pielke et al. (2008) and Fig. 3.7. Globally, there is a trend toward more risk-taking with a growing share of population and assets being located in at-risk areas: between 1970 and 2010, global population grew by 87 %, but the population living in flood plains increased by 114 % and in cyclone prone coastlines by 192 %. The GDP exposed to tropical cyclones increased from 3.6 % of total GDP to 4.3 % over the same period (UN-ISDR 2011).

These results seem at odd with cross-section analyses that suggest that losses are increasing less rapidly than GDP (as stated earlier, Skidmore and Toya find that a 1 % in GDP leads to a 0.5 % increase in disaster losses). One potential explanation is the very high heterogeneity in risk exposure and in risk management trends. To highlight this heterogeneity, one can estimate simple linear regressions of the logarithm of the annual-loss-to-GDP ratio with respect to the logarithm of the GDP per capita, for six countries or regions (Africa, Asia, Europe, North America, South America, Oceania, Japan, US).[5]

[5]Disaster data are from the EM-DAT database for the 1960–2008 period. Countries have been aggregated in large regions to increase the number of disasters included in the analysis. In each region, we aggregate all losses for each year and divide this total loss by the total GDP of the region, to obtain the ratio of total annual loss to total GDP. In the selected regions, there is at least one disaster in most years, limiting the problems linked to null values. When there are years

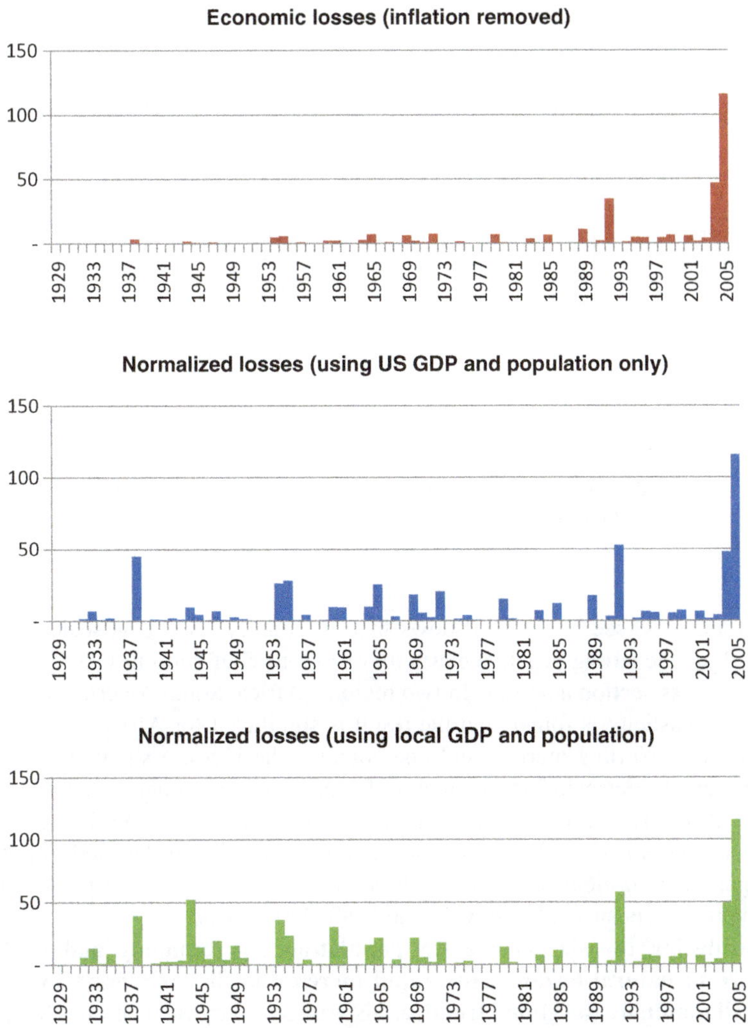


**Fig. 3.7** Hurricane losses in the US from 1929 to 2005. The *top panel* presents the hurricane losses with only the inflation removed; the *middle panel* presents the normalized losses, in which the effect of increasing population and wealth in the US has been removed; the *bottom panel* shows the normalized losses when trends in local GDP and population in hurricane-prone area has been removed. The absence of a trend in the *bottom panel* shows that the increase in hurricane losses in the US is fully explained by socio-economic drivers (Source: Pielke et al. 2008)

---

without disaster, they are removed from the dataset. The number of observations (and thus the number of years without disasters) is indicated in the table.

**Table 3.3** Loss-to-GDP ratio elasticity to GDP per capita for a selected set of countries and regions

| Country or region | Number of observations | Loss-to-GDP ratio elasticity | Error std. | Student't | p value | |
|---|---|---|---|---|---|---|
| Africa | 46 | −1.92532 | 0.962608 | −2.000 | 0.0517 | * |
| Asia | 49 | 1.07195 | 0.290695 | 3.688 | 0.0006 | *** |
| Europe | 45 | 0.589656 | 0.688002 | 0.8571 | 0.3962 | |
| North America | 49 | 1.24228 | 0.681408 | 1.823 | 0.0747 | * |
| South America | 47 | −1.34128 | 1.52525 | −0.8794 | 0.3839 | |
| Oceania | 24 | −2.37486 | 1.82301 | −1.303 | 0.2061 | |
| Japan | 33 | 0.410997 | 0.610413 | 0.6733 | 0.5057 | |
| US (all disasters) | 47 | 1.42383 | 0.640362 | 2.223 | 0.0312 | ** |
| US (hurr., 1929–2005) | 73 | 0.0838976 | 0.476372 | 0.1761 | 0.8607 | |
| US (hurr., 1960–2005) | 46 | 0.350285 | 1.28573 | 0.2724 | 0.7853 | |
| US (hurr., 1980–2005) | 26 | 6.32246 | 2.88914 | 2.188 | 0.0386 | ** |

The last column indicates the significance level, with *** for significance at the 1 % level, ** at the 5 % level, and * at the 10 % level. Unless indicated otherwise, the time period is 1960–2008 (49 years)

This analysis (see Table 3.3) shows that all country- or region-scale analyses do not show the strong negative elasticity of loss-to-GDP ratio to GDP per capita found by cross-section analyses. In two regions (Africa, South America, Oceania), a negative elasticity is found, and the result is significant for Africa. In two other regions (Asia, North America) and one country (the U.S.), a significant positive elasticity can be detected.[6] In one region (Europe) and one country (Japan), a non-significant positive elasticity can be observed. Using only hurricanes in the US and the entire dataset from 1929 to 2005, no significant elasticity is detected. A positive non-significant elasticity is however observable for the 1960–2005 period, and a large significant positive elasticity for the 1980–2005 period.

**Considering the size of the US population and economy and its large exposure to natural hazards, the rapid increase in disaster losses in the US may well dominate the global trend,** as also suggested by Bakkensen (2013) who finds that the US is an outlier in terms of trends in disaster risk.

**Another bias may exist at the international level, since the reporting of natural disaster is better today than it was a few decades ago.** Large-scale floods with significant damages in China were unlikely to be recorded in the 1950s and 1960s, while they are well documented today. This trend toward better information may explain part of the increase in the frequency of disasters, especially in developing countries. This effect is likely to play a continuing role in the future as reporting improves in low-income countries: today, it is still extremely difficult to have good data on disaster impacts, especially for economic losses.

[6] In the two regions (North America and Asia), there is at least one disaster per year, so there is no problem linked to null values. In the U.S., only 2 years have no disaster.

**This is particularly the case for frequent but low-impact events that can be important for population welfare and economic activity but are usually under-reported**: while a huge flood is always reported (and assessed in human and economic terms), many poor countries are affected by very frequent floods. Individually these floods do not cause large damages, but because of the high frequency, the total impact on average annual losses and risk can be significant. Nevertheless, there is no systematic report of these events in most of the developing world.

**In sum, the *current* trends in disaster losses are well explained by socio-economic factors. Climate change does not play a significant role in these evolutions, except maybe for heat waves** (Neumayer and Barthel 2011). **Overall, changes in exposure – that is, the number of people and the value of assets located in at-risk areas – are responsible for the increasing trend in disaster losses that can be observed at global and local scales.**

## 3.4 "Good" and "Bad" Risk-Taking

Disaster economic losses are the cost of risk-taking, and they are on the rise. But why is it so? Maybe in part because locating assets in risky areas also yields benefits, and suffering from increased disaster losses could therefore be a rational decision, based on a trade-off between disaster losses and risk-taking benefits. There is "good" risk-taking that is justified by its benefits. And there is "bad" risk-taking that is linked to bad risk management.

### *3.4.1 Good Risk-Taking*

The "good" reasons to invest in at-risk areas include the following:

- **Investing in at-risk areas can create additional growth from international trade in areas where there are low transportation costs though they are exposed to flood risks**. New Orleans started as a port, and then became an important city that attracted a growing population. This population faced a scarcity of land that forced it to settle in low-lying areas, protected by fragile sea walls and pumping systems.
- **Locating assets in at-risk areas can also increase production when it generates agglomeration externalities**. New York City is located on the coast because it was a major port and it is nowadays one of the largest economic hubs in the world. Today, firms in the finance sector are created in the city not because they benefit from being on the coast and in a flood-prone area, but to benefit from the spill-overs from other industries: (i) existing infrastructure for transportation, water, energy, and communication; (ii) large labor markets with

access to skilled workers; and (iii) access to public services and amenities (art and culture, schools, university, etc.). The same process is at play in developing countries, where the drivers of economic growth are concentrated in cities. Combined with reduced income from agriculture – especially in the poorest countries – these opportunities have created strong incentives for rapid and difficult-to-control rural-urban migrations. Confronted with land scarcity in large cities, these migrations have led to construction in at-risk areas (e.g., Burby et al. 1991, 2001, 2006; Lall and Deichmann 2010). In the most marginal and risky locations, informal settlements and slums are often present, putting a poor and vulnerable population in situations of extreme risk (e.g., Ranger et al. 2011).

– **Investing in at-risk areas can also increase utility through environmental amenities** (e.g., from sea views) and generate revenues from tourism. Current trends in disaster losses could thus be – at least partly – justified by non-observed benefits from increasing risk-taking.

### *3.4.2 Bad Risk-Taking*

The fact that *some* risk-taking is justified by the benefits it yields does not mean that *all* risk-taking is good, that current levels of risk-taking are optimal or that current trends are sustainable or desirable. They are also many reasons why "bad" risk-taking is taking place:

- **Limited resources and choice limitation:** many households have little freedom in their localization choices, especially in crowded mega-cities, and they therefore settle in at-risk areas in lack of alternatives. Also, households face financial constraints that make risk management impossible. For example, flood- and earthquake-proofed homes costs $3,100 more than a typical home in Jakarta. This is only slightly less than the annual per capita GDP in Indonesia and most households cannot afford what may appear to them as a luxury.
- **Distorted incentives, externalities, moral hazards, and market failures**: since insurance and post-disaster support are often available in developed countries, households and firms in risky areas do not pay the full cost of the risk, and may take more risk than what is optimal socially (e.g., Kaplow 1991; Burby 2006; Laffont 1995). Also, some actors take risks and benefit from them, while other individuals bear the risk and costs if the risk materializes: when a developer builds housing in a flood plain and sells to ill-informed customers, he or she gets the benefits from the increase in risk while transferring the potential losses to others. In that case, the developer has little self-interest in canceling his or her project or in reducing his or her profit with expensive risk mitigation investments.
- **Information and transaction costs**: since information on natural hazards and risk are not always easily available, households and businesses may decide not to spend the time, money and effort to collect them, and disregard this information in their decision-making process. (Magat et al. 1987; Camerer and

Kunreuther 1989; Hogarth and Kunreuther 1995). For example, only 31 % of people living in areas prone to flooding in the United States are aware of this risk, and only 33 % know that federally backed flood insurance was available. Also, data is sometimes expensive, making it difficult for some actors to get this information and act accordingly. For instance, the cost of one meteorological observation in Europe varies from zero (in Slovenia) to 0.40 euros (in Romania); for operational risk management, long data series are needed and the total cost can be discouraging for some actors.

- **Irrational behaviors and biased risk perceptions:** Individuals do not always react rationally when confronted with small probability risks (Magat et al. 1987; Camerer and Kunreuther 1989; Hogarth and Kunreuther 1995), and they defer choosing between ambiguous choices (Tversky and Shafir 1992; Trope and Liberman 2003). Moreover, they have trouble taking into account events that have never occurred before (the "bias of imaginability", see Tversky and Kahneman 1974). Finally, private and public investment decisions do not always take adequately long or very long-term consequences into account and people have short memories for disasters (for public decisions, see Michel-Kerjan 2008; for private decisions, see Kunreuther and Slovic 1978; Kunreuther et al. 1978; Thaler 1999). After Hurricane Katrina hit the US coast in 2005, the number of U.S. households with flood risk insurance increased more than three times more rapidly than observed in previous years. But the average cancellation rate remained unchanged, at approximately 33 % a year, suggesting a short effect of the disaster on long-term household behavior. There is also a "paradox of protection": when protection against frequent disasters cancels losses for a long period of time, vigilance and risk awareness decrease. That leads to lack of maintenance of protective infrastructure and increased investments in risky areas, resulting in higher future losses (and losses of increasing scale).
- **Political economy obstacles:** decision-makers are reluctant to invest in risk management because the costs of risk management are immediate, concentrated, and observable, while the benefits are longer term, widely distributed, and less visible. When prohibiting or regulating the development of a flood area, for instance, decision makers have a strong and immediate influence on land values. Affected people will oppose the constraint and be very vocal about it. On the other hand, the people who benefit from the regulation – e.g., future buyers of apartments in the newly developed areas – are often not aware that the regulation will protect them and therefore do not support it publicly.

### *3.4.3 Consequences of Risk Management Policies*

**The key to success in risk management is to distinguish good risk-taking from bad risk-taking, to support the former and prevent the latter. Such an approach to risk management is not only an instrument to reduce losses from disasters. It is also an instrument that can make it possible for individuals and firms to take**

**risks when benefits exceeds costs, and thus to capture opportunities.** As such, risk management can be considered as an "economic growth instrument", which can help individuals and firms take the risks that are needed to escape poverty, grow their income and wealth, and improve their well-being.

For instance:

- If farmers can get access to drought insurance, they are more able to borrow the funds necessary for buying fertilizers, tools and other supplies, therefore increasing their average productivity. In the absence of insurance (or other post-drought support), the possibility of greatly reduced production due to destructive weather conditions – and thus of being unable to repay loans– deters farmers from investments that are on average very productive.
- Households could invest in better housing quality – especially in slums and informal settlements – but the risk of losing everything in case of such disasters as floods may deter them from doing so. Providing a post-disaster support instrument, and efficient banking instruments to help them save in monetary forms, can assure their ability to deal with disasters if and when they occur, and can make it possible for them to invest in home improvement (e.g., toilets, drinking water) that can vastly improve their welfare and productivity.

**The existence of a risk can have negative consequences on development, growth, and poverty reduction, even if the risk does not materialize into a disaster.** If households know that they may lose their assets anytime, they might decide not to accumulate them and to consume their income instead (or to keep their savings in non-productive forms). This would have a negative impact on their future income and thus on their ability to escape poverty.

To help design risk management policies that can separate positive from negative risk-taking, the 2014 World Development Report proposes to use a series of analyses, designed to identify situations in which risk-taking would be insufficient or excessive, and to design policies to correct these problems.

In this framework, designing effective public policy for risk management starts with an assessment of existing risk, but goes way beyond it. Next, it analyzes the reasons why risk management is not optimal, through a sequence of assessments (Fig. 3.8):

- Risk management may be inefficient because decision makers (private or public) face incentives that are distorted toward too much or too little risk-taking. If this is the case, then this problem needs to be fixed first, since resources and information are unlikely to be used wisely in the present of distorted incentives.
- If incentives are correct, risk management can be inefficient because decision-makers do not have the information they need to translate incentives into productive actions.
- If incentives are correct, information is accessible, but risk management is still inefficient, then it might be because of cognitive and behavioral issues (e.g., the information exists but not in a form that can be processed by decision-makers, for instance because it is too complex or too abstract).

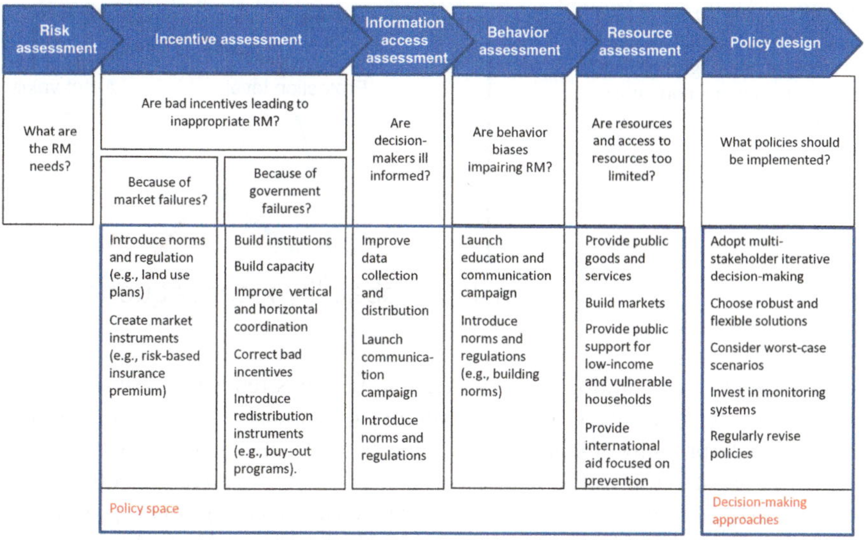


**Fig. 3.8** A set of screens for assessing obstacles to risk management, and formulating policy responses (Source: World Bank 2013)

- Finally, if decision-makers are willing to improve their risk management, but cannot do it because of lack of resources, then the provision of additional resources is helpful.

For each of these problems, Fig. 3.8 also provides a series of options to improve the situation, such as a communication campaign if the problem is related to information access or behaviors.

## 3.5 Technical Insight: Economic Growth and Disaster Losses

This section proposes a small model to get a better understanding of the link between development, economic growth, and risk.

### *3.5.1 Risk and Development*

Economic growth affects the average level of disaster losses (i.e., the risk level) through two main channels. First, higher income makes it possible to invest in better protections against disasters, therefore reducing the probability of a disaster (a *protection effect*). But higher income is also likely to increase the value of assets located in risky areas, and therefore to increase the consequences if a disaster occurs

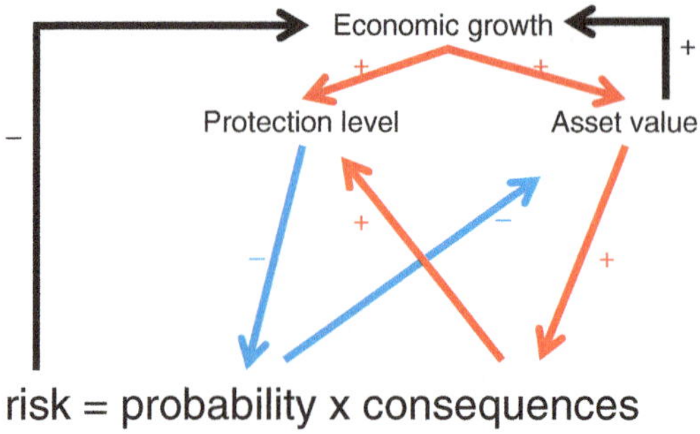


**Fig. 3.9** A simple risk framework to analyze the link between economic growth and risk-taking in a normative setting

(a *wealth effect*). In addition, some protection measures reduce the consequences of a disaster without affecting the probability: for instance, reinforced buildings (e.g., wet waterproofing of houses) can reduce the damages to buildings in case of floods without affecting the probability of such an event (a *vulnerability* effect).

The impact of economic growth on risk – defined as the product of the probability and the consequences – is therefore ambiguous. And the system is made more complex by the interactions between these two channels: more protection reduces the disincentive to invest in risky areas and can magnify the wealth effect. And higher asset value in at-risk areas increases the incentive to protect the area, magnifying the protection effect. Overall, the decisions concerning protection and risk-taking are also likely to feedback on economic growth, for instance because coastal development has been revealed to be critical for export-led growth. Figure 3.9 summarizes these different potential effects and their interactions (assuming that protection only decreases the probability of a disaster, not its consequences).

### 3.5.2 A Balanced Growth Pathway

Let us assume a local economy, in which economic production is supported by productive capital only, with decreasing returns:

$$Y_b = f\left(K_b\right) = \phi K_b^{\varphi}. \tag{3.2}$$

The variable $Y_b$ is annual production (i.e. value added) in the balanced growth pathway; $K_b$ is the corresponding amount of productive capital; $\phi$ is total productivity; and $\varphi$ is a parameter that describes the decreasing returns of capital ($\varphi < 1$). All variables are time dependent, and are assumed to growth over time. Productivity is growing at a rate λ.

$$\phi(t) = \phi(0)e^{\lambda t}. \tag{3.3}$$

Assuming the economy is on a balanced growth pathway, production and capital are growing at the same rate:

$$Y_b(t) = Y_b(0)e^{\mu t}, \tag{3.4}$$

$$K_b(t) = K_b(0)e^{\mu t}. \tag{3.5}$$

To be consistent, Eqs. (3.2, 3.3, 3.4 and 3.5) require:

$$\mu = \frac{\lambda}{1-\phi}, \tag{3.6}$$

and

$$Y_b(0) = \phi(0)(K_b(0))^{\varphi}. \tag{3.7}$$

The variable $\mu$ is the economic growth rate. Based on this simple model, we can now introduce risk taking and protection investments.

### 3.5.3 *The Safety vs. Productivity Trade-Off*

Let us now assume that the amount of capital $K_b$ can be either located in safe locations ($K_s$) or in risky locations ($K_r$), with $K_b = K_s + K_r$. Examples of risky locations are coastal areas, where storm surge and coastal floods are possible, but also areas at risk of river floods, high concentration urban areas at risk of floods in case of heavy precipitations, and even earthquake and hurricane-prone regions. As is observed in the real world, we assume that risky locations are more productive, thanks to their location (e.g., proximity of port infrastructure for export-oriented industries; coastline amenities for tourism; easier access to jobs in at-risk locations in crowded cities). This increase in production has decreasing returns, however. As a consequence, total production becomes:

$$Y = Y_b + \Delta Y = Y_b + \alpha K_r^{\gamma}, \tag{3.8}$$

where $\Delta Y$ is the additional output produced thanks to the localization of capital in risk-prone areas; $\alpha$ is a relative productivity advantage, and is assumed to growth at the same rate than the general productivity $\phi$ (i.e., at the rate $\lambda$).

Real capital located in a risky area can be affected by hazards, like floods and windstorms. If a hazard is too strong, it causes damages to the real capital installed in at-risk areas, and can be labeled as a disaster. To simplify the analysis, we assume that in that case, the real capital at-risk is totally destroyed. It is assumed that this is the only consequence of disasters; no other impacts – like fatalities and casualties – are considered in this simple model. Indirect losses are also neglected.[7]

These disasters (i.e. hazards that lead to real capital destruction) have a probability $p_0$ to occur every year, except if protection investments are carried out and reduce this probability. These protection investments take many forms, depending on which hazard is considered. Flood protections include dikes and seawalls, but also drainage systems to cope with heavy precipitations in urban areas. Windstorm and earthquake protections consist mainly in building retrofits and stricter building norms, to ensure old and new buildings can resist stronger winds or larger earthquakes.

It is assumed that better defenses reduce the probability of disasters, but do not reduce their consequences. This is consistent with many types of defenses, like seawalls that can protect an area up to a design standard of protection but fail totally if this standard is exceeded. Better defenses are also more expensive, and the annual cost of defenses C and the remaining disaster probability p are assumed to be linked by the relationship:

$$C(p) = \xi \left( \frac{1}{p^{\nu}} - \frac{1}{p_0^{\nu}} \right). \tag{3.9}$$

The cost of reducing flood probability to zero is infinite. Depending on the value of $\nu$, protection costs increase more or less rapidly when the disaster probability approaches zero. The parameter $\nu$ corresponds therefore to more or less optimistic assumptions on protection costs. Any given year, the economic output is given by:

$$Y = Y_b + \alpha K_r^{\gamma} - C(p) - L, \tag{3.10}$$

where L is damages from disasters, and is given by a random draw with probability p. If a disaster occurs, losses are equal to $K_r$, i.e. all the real capital located in the risky area is destroyed. Any given year, the expected loss E(L) is equal to $pK_r$ and the expected output is equal to:

$$E(Y) = Y_b + \alpha K_r^{\gamma} - C(p) - pK_r. \tag{3.11}$$

[7]If indirect losses depend linearly on direct losses, then the analysis remains unchanged. If they are nonlinearly increasing with direct losses (as modeling analysis suggests, see Sect. 2.2.3), then it affects model results and the model can no longer be solved analytically.

### 3.5.4 Optimal Protection and Risk-Taking

Assuming a social planner decides on the amount of capital $K_r$ to be located in the risky area and of the level of protection that is to be built, his program is[8]:

$$\max_{p,\,K_r} E\,(Y)\,, \tag{3.12}$$

under the constraints that $K_r \leq K_b$ and $0 \leq p \leq p_0$.

**We assume first that there is no risk aversion and we assume that the expected output is maximized.** From the social planner's perspective, doing so is acceptable if disaster losses remain small compared to aggregated income, consistently with the Arrow-Lind theorem for public investment decisions (Arrow and Lind 1970). As discussed in Mahul and Ghesquiere (2007), this theorem holds only if some conditions are met, including if disaster losses can be pooled among a large enough population (e.g., a large country), and with many other uncorrelated risks, i.e. in the presence of comprehensive insurance coverage or post-disaster government support, or if disaster losses can be smoothed over time thanks to savings and borrowing (i.e. self-insurance) or reinsurance. In other terms, the optimal pathways determined by this analysis are valid assuming that the social planner ensures that individual losses remain small thanks to temporal smoothing and redistribution or insurance across individuals. **In particular, it means that human losses (fatalities and casualties) can be avoided for instance thanks to early warning and evacuation.**

First order conditions lead to the equalities:

$$\overline{p} = (\nu\xi)^{\frac{\gamma-1}{\gamma+\nu(\gamma-1)}}(\alpha\gamma)^{\frac{1}{\gamma+\nu(\gamma-1)}}, \tag{3.13}$$

$$\overline{K_r} = (\nu\xi)^{\frac{1}{\gamma+\nu(\gamma-1)}}(\alpha\gamma)^{-\frac{\nu+1}{\gamma+\nu(\gamma-1)}}. \tag{3.14}$$

Assuming that $Kr \leq Kb$ and that $p \leq p_0$, the expected annual loss is equal to

$$\overline{E(L)} = (\nu\xi)^{\frac{\gamma}{\gamma+\nu(\gamma-1)}}(\alpha\gamma)^{-\frac{\nu}{\gamma+\nu(\gamma-1)}}, \tag{3.15}$$

and the loss in case of flood is equal to

$$\overline{L} = (\nu\xi)^{\frac{1}{\gamma+\nu(\gamma-1)}}(\alpha\gamma)^{-\frac{\nu+1}{\gamma+\nu(\gamma-1)}}. \tag{3.16}$$

[8]This model is different from the Schumacher and Strobl (2011) model. In the latter, the only decision concerns protection investments that mitigate disaster consequences, and there is no benefit from taking risks and thus no trade-off between safety and higher income.

When productivity $\alpha$ is growing over time at the rate $\lambda$, there are two possibilities, depending on the value of $\gamma$, the exponent representing decreasing returns in the additional productivity from capital located in at-risk areas; see Eq. (3.8).

If $\gamma > \nu/(\nu+1)$, then $\overline{K}_r$ and $\overline{E(L)}$ are decreasing over time *in absolute terms*. In that case, less and less real capital is installed in the risk area when productivity and wealth increases. So, the absolute level of risk is decreasing with wealth. It is also noteworthy that, in such a situation, annual mean losses and capital at risk counter-intuitively decrease if protection costs ($\xi$) increase.

If $\gamma < \nu/(\nu+1)$, then the amount of capital at-risk increases, and the risk (both in terms of average loss and maximum loss) is increasing with wealth, and mean annual losses and capital at risk are augmented if protection costs ($\xi$) increase.

But the absolute level of risk is not a good measure of risk: a wealthier society is able to cope with larger losses. The question is therefore the relative change in risk. One way of investigating this question is to assess whether $\overline{K}_r$ and $\overline{E(L)}$ are growing more or less rapidly than $Y_b$ and $K_b$, i.e. at a rate larger or lower than $\mu$.

If productivity $\alpha$ is growing at a rate $\lambda$, expected losses $\overline{E(L)}$ are growing at a rate $\lambda\frac{-\nu}{\gamma+\nu(\gamma-1)}$ and maximum losses (i.e., the losses in case a flood occurs), $\overline{K}_r$, are growing at a rate $\lambda\frac{-(\nu+1)}{\gamma+\nu(\gamma-1)}$.

Since $K$ and $Y$ are growing at a rate $\mu = \lambda/(1-\phi)$, we have the following result:

If $\phi\frac{\nu}{\nu+1} < \gamma < \frac{\nu}{\nu+1}$, then mean annual losses $\overline{E(L)}$ are growing faster than baseline economic output $Y_b$. If $\phi - \frac{1}{\nu+1} < \gamma < \frac{\nu}{\nu+1}$, then the capital at risk and the losses in case of disasters (i.e. $\overline{K}_r$) are growing faster than $Y_b$.

With usual values for $\phi$, i.e. about 1/3, and the simplest assumption for protection cost, i.e. $\nu = 1$, losses in case of disasters are growing faster than $Y$ for any $\gamma$, positive and lower than 1/2. Mean annual losses increase faster than $Y_b$ if $\gamma$ is between 1/6 and 1/2.

**Therefore, it can be optimum for disaster losses and mean annual losses to increase with wealth in the future, *even in relative terms*. In other terms, the benefits from increasing risk-taking may theoretically exceed the costs, showing that increase disaster losses are not always a signal that there is a problem.**

In this case, all real capital will eventually be installed in at-risk areas ($\overline{K}_r = K_b$), and a disaster would lead to the complete destruction of all production capacities, with a probability $\overline{p} = \left(\frac{\xi\nu}{K_b}\right)^{\frac{1}{1+\nu}}$.

Surprisingly, the increase in risk happens when $\gamma$ is small enough, i.e. when additional productivity from locating real capital in at-risk areas exhibits sufficiently diminishing returns. Consistent with intuition, however, is the fact that increase in risk is more likely when $\nu$ is large, i.e. when protection costs are increasing rapidly with the protection level. It is interesting to note that absolute protection costs ($\xi$) and the absolute additional productivity ($\alpha$) do not influence the behavior of mean annual losses and capital at risk, but only their levels.

If $\gamma = \nu/(\nu+1)$, there is no inside maximum in Eq. (3.12). Instead, there are two possibilities depending on the protection cost relative to the additional productivity in at-risk areas. If the additional productivity is high enough (relative to protection costs), then all real capital is located in at-risk area ($\overline{K}_r = K_b$). If the additional production is not sufficient, then no protection is provided ($\overline{p} = p_0$).

The limit between these two possibilities depends on the protection costs, relative to the additional productivity in at-risk areas. The limit protection cost ($\xi_l$) can be written as a function of the additional productivity $\alpha$ as:

$$\xi_l\left(\alpha\right) = \left(\frac{\alpha}{1+\nu}\right)^{(1+\nu)} \nu^{\nu} \tag{3.17}$$

and – equivalently – the limit additional productivity can be written as a function of protection costs:

$$\alpha_l\left(\xi\right) = \left(1+\nu\right) c^{-\frac{\nu}{1+\nu}} \xi^{\frac{1}{1+\nu}}. \tag{3.18}$$

If $\gamma = \nu/(\nu+1)$ and $\xi > \xi_l(\alpha)$ (or, equivalently, $\alpha < \alpha_l(\xi)$), then no protection is provided ($\overline{p} = p_0$) and the capital in at-risk areas is equal to $\overline{K}_r = \left(\frac{\alpha\nu}{p_0(1+\nu)}\right)^{(1+\nu)}$. If $\gamma = \nu/(\nu+1)$ and $\xi < \xi_l(\alpha)$ (or, equivalently, $\alpha > \alpha_l(\xi)$), then all capital is located in at-risk areas (i.e., $\overline{K}_r = K_b$), and the protection is such that the flood probability is equal to $\overline{p} = \left(\frac{\xi\nu}{K_b}\right)^{\frac{1}{1+\nu}}$.

## 3.6 Conclusion

**This chapter proposes an economic framework to analyze the trade-off between disaster losses and the benefits from taking risks.** It shows that there is a potential negative side-effect from higher protection against disaster, in the form of an increasing vulnerability to exceptional events. Protection reduces the hazard (the probability of occurrence of an event), but its impact on risk is more complex, because it also increases exposure (here, the capital at risk $K_r$). As a result, protection transfers part of the risk from one kind of risk (frequent and low-cost events) to another kind (exceptional and high-impact events), a process already stressed in Etkin (1999).

**In some cases, this negative side-effect leads to an increase in risk level, in spite of continuously improving protection.** A model suggests that the overall risk – i.e. mean annual losses – can increase with time in spite of a decreasing disaster probability. The risk can even increase faster than wealth in some circumstances.

**Disaster risk management policies need to acknowledge the benefits from investing in at-risk areas, and take these benefits into account in their design.** If individuals and businesses take risks that are inappropriate at the social scale – for instance because they cannot deal with a bad outcome and will then get support from the government and taxpayers – then the best course of action is different depending on what explains this behavior.

**If risk-taking is due to irrational behaviors and imperfect information only, then communication tools and land-use regulations are easy to implement and should be able to reduce the level of risk.**

**But if risk-taking is justified by the benefits derived from investing and settling in at-risk areas, then efficient risk management policies will fail or be economically detrimental if they do not provide alternatives to get similar benefits without taking those risks.** If investments in risky coastal areas are explained by the benefits from the proximity of a port, for instance, then risk can be reduced by providing businesses with safe development areas that are connected to the port by efficient transport infrastructure. And when activities require proximity to the port and no alternative exists, then investment should be allowed in at-risk areas, provided that (a) social benefits justify it, and (b) specific risk mitigation actions are undertaken (from warning and evacuation systems to compulsory insurance with risk-based premium). Similarly, if newcomers settle in risky areas of mega-cities because it is the only way to have access to the jobs and opportunities offered by the city, then risk cannot be reduced by simply prohibiting settlements in these areas. Such a policy would face strong political opposition and, if implemented, might lead to the creation of informal settlements, thereby increasing the level of risk instead of reducing it. An efficient risk management policy should rather propose viable alternatives to newcomers, for instance by providing cheap and rapid (possibly subsidized) public transportation from job centers to safe areas that can be developed.

The typical approach for risk mitigation is based on "flood zoning," i.e. on the definition of flood-prone areas where investment and settlement are prohibited. This approach, however, faces difficulties in implementation and enforcement, because it neither distinguishes among different types of economic activities nor accounts for potential benefits from at-risk investments. A more flexible approach that accounts for these benefits – through economic analysis or consultative processes – is more likely to be effective in reducing risks and to provide net economic benefits.

**More generally, disaster risk management policies need to shift from a purely negative stance – indicating where it is prohibited to invest and settle – to a more positive approach – indicating where investments should be directed, and providing complementary measures that can make these investments as beneficial as those in at-risk areas.** To do so, risk management should not focus only on at-risk areas, but follow a more holistic approach, integrated in development planning. Sometimes, building a transportation infrastructure to connect job centers to safe housing areas is more efficient in reducing risk levels than building dikes to protect a flood-prone zone.

## References


Arrow KJ, Lind RC (1970) Uncertainty and the evaluation of public investment decisions. Am Econ Rev 60:364–378

Bakkensen LA (2013) Adaptation and natural disasters: evidence from global tropical cyclone damages and fatalities. In: EAERE 2013 annual conference, Toulouse

Barredo JI (2009) Normalised flood losses in Europe: 1970–2006. Nat Hazards Earth Syst Sci 9:97–104

Burby JR (2006) Hurricane Katrina and the paradoxes of government disaster policy: bringing about wise governmental decisions for hazardous areas. Ann Am Acad Polit Soc Sci 604(1):171–191

Burby RJ, Cigler BA, French SP, Kaiser EJ, Kartez J, Roenigk D, Weist D, Whittington D (1991) Sharing environmental risks: how to control governments' losses in natural disasters. Westview, Boulder

Burby RJ, Nelson A, Parker D, Handmer J (2001) Urban containment policy and exposure to natural hazards: is there a connection? J Environ Plan Manag 44(4):475–490

Burby RJ, Nelson AC, Sanchez TW (2006) The problem of containment and the promise of planning. In: Birch EL, Wachter SM (eds) Rebuilding urban places after disaster, lessons from Hurricane Katrina. University of Pennsylvania Press, Philadelphia

Camerer C, Kunreuther H (1989) Decision processes for low probability events: policy implications. J Policy Anal Manag 8:565–592

Etkin D (1999) Risk transference and related trends: driving forces towards more mega-disasters. Environ Hazards 1:69–75

Hogarth R, Kunreuther H (1995) Decision making under ignorance: arguing with yourself. J Risk Uncertain 10:15–36

IPCC (2012) Managing the risks of extreme events and disasters to advance climate change adaptation. A special report of Working Groups I and II of the Intergovernmental Panel on Climate Change [Field CB, Barros V, Stocker TF, Qin D, Dokken DJ, Ebi KL, Mastrandrea MD, Mach KJ, Plattner G-K, Allen SK, Tignor M, Midgley PM (eds)]. Cambridge University Press, Cambridge/New York, 582 pp

Jonkman SN, Bockarjova M, Kok M, Bernardini P (2008) Integrated hydrodynamic and economic modelling of flood damage in the Netherlands. Ecol Econ 66:77–90

Kahn ME (2005) The death toll from natural disasters: the role of income, geography, and institutions. Rev Econ Stat 87(2):271–284

Kaplow L (1991) Incentives and government relief for risk. J Risk Uncertain 4(2):167–175

Kellenberg DK, Mobarak AM (2008) Does rising income increase or decrease damage risk from natural disasters? J Urban Econ 63:788–802

Kunreuther H, Slovic P (1978) Economics, psychology, and protective behavior. Am Econ Rev 68(2). Papers and proceedings of the ninetieth annual meeting of the American Economic Association, pp 64–69

Kunreuther H, Ginsberg R, Miller L, Sagi P, Slovic P, Borkan B, Katz N (1978) Disaster insurance protection: public policy lessons. Wiley, New York

Laffont JJ (1995) Regulation, moral hazard and insurance of environmental risks. J Public Econ 58(3):319–336

Lall SV, Deichmann U (2010) Density and disasters: economics of urban hazard risk, Policy Research working paper 5161. The World Bank, Washington, DC

Magat W, Viscusi KW, Huber J (1987) Risk-dollar tradeoffs, risk perceptions, and consumer behaviour. In: Viscusi W, Magat W (eds) Learning about risk. Harvard University Press, Cambridge, pp 83–97

Mahul O, Ghesquiere F (2007) Sovereign natural disaster insurance for developing countries: a paradigm shift in catastrophe risk financing. World Bank Policy Research working paper 4345. Washington, DC, USA

Michel-Kerjan E (2008) Disasters and public policy: can market lessons help address government failures. In: Proceedings of the 99th National Tax Association conference, Boston

Miller S, Muir-Wood R, Boissonnade A (2008) An exploration of trends in normalized weather related catastrophe losses. In: Diaz HF, Murnane RJ (eds) Climate extremes and society. Cambridge University Press, Cambridge, pp 225–347

Neumayer E, Barthel F (2011) Normalizing economic loss from natural disasters: a global analysis. Glob Environ Chang 21(1):13–24

Nordhaus WD (2006) The economics of hurricanes in the United States, prepared for the Snowmass workshop on abrupt and catastrophic climate change. Annual Meetings of the American Economic Association, Snowmass

Peduzzi P, Chatenoux B, Dao H, Herold C, Giuliani G (2011) Preview global risk data platform. UNEP/GRID and UNISDR, Geneva, Switzerland. preview.grid.unep.ch/index.php?preview=tools&cat=1&lang=eng

Pielke RA Jr, Gratz J, Landsea CW, Collins D, Saunders MA, Musulin R (2008) Normalized hurricane damages in the United States: 1900–2005. Nat Hazards Rev 9(1):29–42

Ranger N, Hallegatte S, Bhattacharya S, Bachu M, Priya S, Dhore K, Rafique F, Mathur P, Naville N, Henriet F, Herweijer C, Pohit S, Corfee-Morlot J (2011) A preliminary assessment of the potential impact of climate change on flood risk in Mumbai. Clim Chang 104(1):139–167

Schumacher I, Strobl E (2011) Economic development and losses due to natural disasters: the role of risk. Ecol Econ 72:97–105

Skidmore M, Toya H (2007) Economic development and the impacts of natural disasters. Econ Lett 94:20–25

Thaler R (1999) Mental accounting matters. J Behav Decis Mak 12:183–206

Trope Y, Liberman N (2003) Temporal construal. Psychol Rev 110(3):403–421

Tversky A, Kahneman D (1974) Judgment under uncertainty: heuristics and biases. Science, New Series 185(4157):1124–1131

Tversky A, Shafir E (1992) Choice under conflict: the dynamics of deferred decision. Psychol Sci 3(6):358–361

UN-ISDR (2011) Revealing risk, redefining development. United Nations international strategy for natural disaster reduction global assessment report on disaster risk reduction. Geneva, Switzerland, 178 pp

World Bank (2011) Climate change adaptation and natural disasters preparedness in the coastal cities of North Africa. The World Bank, Washington, DC

World Bank (2013) World development report 2014. Risk and opportunity: managing risk for development. World Bank, Washington, DC

# Chapter 4
# Trends in Hazards and the Role of Climate Change

**Abstract** This chapter reviews how climate change may affect natural hazards in the future, stressing the heterogeneity in its effect on hazard depending on the type of hazard (e.g., heavy rainfall vs. snow storm) and the region of the world. It concludes that the recent growth in disaster economic losses is overwhelmingly due to socioeconomic trends, but that climate change will play an increasing role over time. Analyses suggest that climate change can increase dramatically natural risks in some contexts, especially where protections have been designed for a stable climate and may quickly become ill-adapted to new conditions. More than average losses, the increase in the likelihood of the more intense hazards is a serious reason for concern.

**Keywords** Climate change • Natural hazard • Natural disasters • Disaster losses

Even though exposure and vulnerability are clearly the dominant drivers of current trends in disaster losses (Chap. 3), it does not mean that climate change will not play an important role in the future. This chapter summarizes the potential impacts of global warming on extreme events and hazards.

## 4.1 Scenarios for Climate Change Analysis

Investigating climate change requires projecting how human emissions of greenhouse gases (GHG) will evolve in the future. Of course, this evolution depends on many drivers that are impossible to predict, including demographic changes, economic growth, technological changes, and cultural choices. It means that climate change and its impacts cannot be predicted, in the classical sense.

Instead, most climate change impact analyses have been carried out using *scenarios*. These analyses start from one or several baseline scenarios (often from the Special Report on Emission Scenarios or SRES, Nakicenovic and Swart 2000). These baseline scenarios build on different possible pathways of the world-evolution main drivers (population growth, technological change, etc.) to create "possible and consistent" world evolutions, assuming no impact from climate change and no

 
S. Hallegatte, *Natural Disasters and Climate Change*,
DOI 10.1007/978-3-319-08933-1_4

climate policy. Depending on these hypotheses, the scenarios include different levels of greenhouse gas (GHG) emissions, which lead to different amplitudes and patterns of climate change. They also include different socio-economic evolution, which can be translated into different vulnerabilities to climate change. From the resulting change in climate and these vulnerabilities, many studies have then estimated how climate change would affect ecosystems, economies and societies.

**These scenarios are not supposed to be predictions of what the future will look like (we know we cannot do that). But they represent theoretical futures that are possible and internally consistent. The idea is that analyzing climate change impacts in these multiple scenarios will give us a good sense of how dangerous climate change is, even if we cannot predict future impacts. As a result, the result of these analyses is a useful input for decision-making, but cannot be used directly to make a decision (see a discussion in Chaps. 6 and 7).**

The last IPCC report (IPCC 2013) bases its assessment on the use of a set of four scenarios, referred to as the "Radiative Forcing Pathways" or RCPs. They describe the future impact of man-made emissions (and changes in land use) on the climate system (see Fig. 4.1); see Van Vuuren et al. (2011). These scenarios are expressed in terms of globally-averaged "radiative forcing", i.e. the change in the energy flux at the top of the atmosphere due to the changes in atmosphere composition (e.g., $CO_2$ concentration), and they do not include socio-economic information.

At the equilibrium of the earth energy balance, there is no net flux at the top of the atmosphere. With increasing concentration of greenhouse gases, however, the energy budget of the planet is perturbed and there is a positive net downward flux of energy that increases the planet's average temperature (until the increase in temperature is sufficient to balance the energy budget and bring the radiative forcing to zero in spite of higher GHG concentrations). This flux, measured before the atmosphere temperature had increased, is what is referred to as radiative forcing.[1]

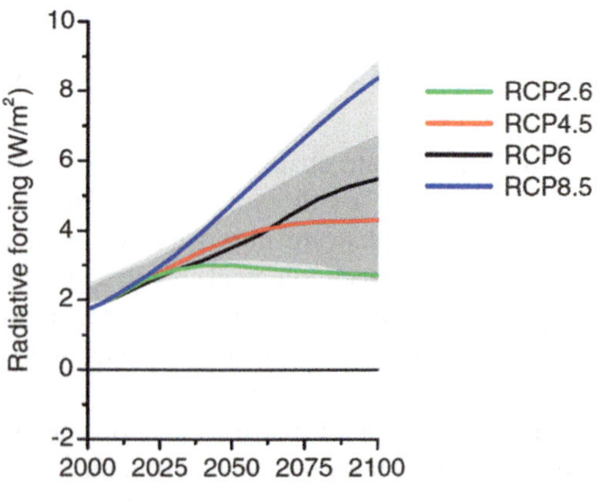


**Fig. 4.1** The four scenarios used in the IPCC (2013) to characterize the impact of manmade emissions on the climate. The four scenarios represent different levels of "radiative forcing" (equivalent to an additional flux of energy at the top of the atmosphere) (From Van Vuuren et al. 2011)

[1]To be precise, the radiative forcing is the flux when the stratosphere (i.e. the upper-atmosphere) has reached its radiative equilibrium, which is extremely rapid (a few days), and before the troposphere (i.e. the lower atmosphere, below 10 km on average) could adjust.

Overall, radiative forcing is a good index of the human influence on the climate, and it is a convenient measure as it combines the effect of the many greenhouse gases that are emitted by human activities.

To serve as inputs for climate models, each RCP includes a pathway for the global radiative forcing due to all long-lived gases ($CO_2$, methane, etc.), but also (i) a spatialized scenario for land-use (that describes which areas are covered by forest, agriculture, or artificial soil), and (ii) a spatialized scenario for short-lived gases (e.g., ozone) that also have an impact on local and global climates. Short-lived gases cannot be included in the radiative forcing because their impact is not homogeneous and their spatial distribution depends on where they are emitted.

These scenarios do not include socio-economic information such as population change, economic development, urbanization, etc. So the RCPs cannot be used alone to assess the impact of climate change on societies and economies. They need to be complemented with information on social and economic vulnerability.

Using scenarios of radiative forcing and other forcing, however, climate modelers can run their climate models and assess the physical consequences of human activities on the climate system without considering the drivers of their forcing.[2]

## 4.2 Climate Change Scenarios

### *4.2.1 Changes in Average Climate Conditions*

From each of these RCP scenarios, climate models (or Atmosphere-ocean General Circulation Models, AOGCMs or sometimes GCMs) can produce climate scenarios, i.e., possible evolutions of all meteorological variables (temperature, precipitation, wind, etc.). These models project that global temperatures will increase along this century, by between 1 and 5°C, depending both on future emission levels, and on the climate model that is used (see Fig. 4.2).

All models show roughly the same pattern of warming. Figure 4.3 shows this pattern for the average model (i.e., the average of all available models), for 1°C of global warming. The right-hand side panel shows the average temperature change as a function of the latitude. This pattern shows that the warming will be larger on continents than on the ocean and larger in the higher latitudes than in the tropics. It is important to note that a +1°C increase in global temperature leads to temperature increases of more than 2°C in northern Polar Regions. This map shows that climate change, and thus its impacts, will be very heterogeneous.

Of course, climate change is not only about temperature. Precipitation patterns also will be modified, even though there is much more uncertainty on precipitation

[2]The approach was completely different in the Fourth Assessment Report published in 2007. In this previous report, the IPCC used the SRES scenarios, which were developed in a more linear fashion, starting with scenarios for the evolution of population, technologies, and economies, and then a modeling of resulting emissions and land-use changes.

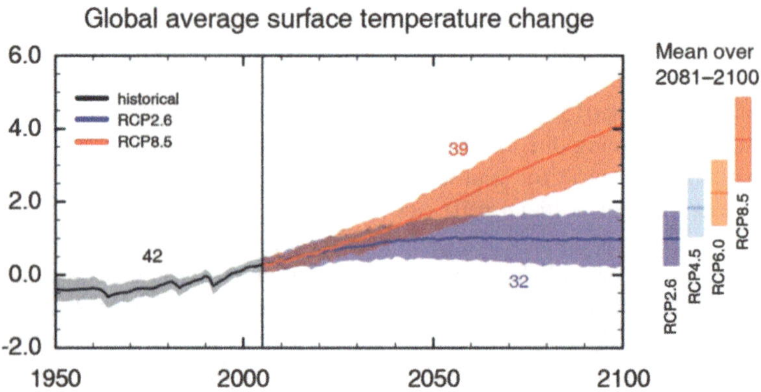


**Fig. 4.2** For different GHG emission scenarios (the four RCP scenarios), climate models can derive climate scenarios, here the corresponding increase in global temperature for the RCP2.6 and RCP8.5 scenarios (From IPCC 2013)

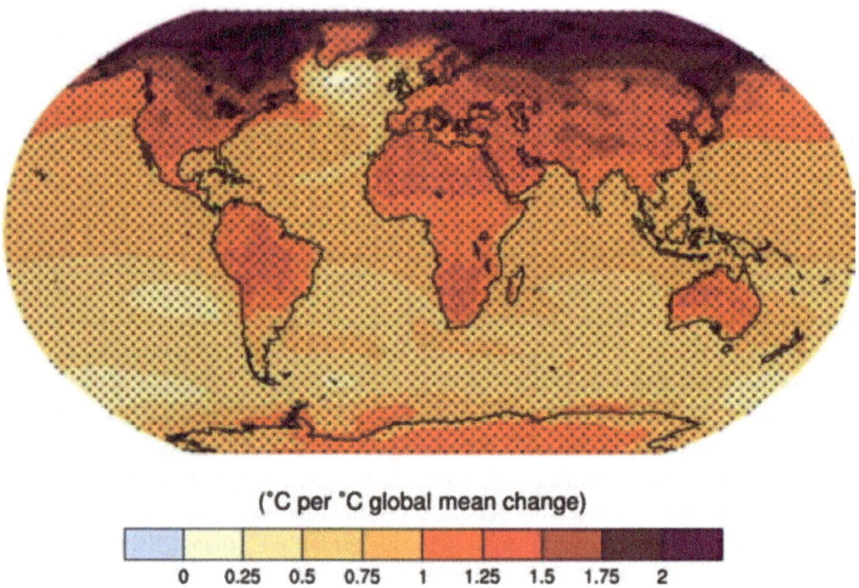


**Fig. 4.3** Geographical pattern of warming, for the average model (i.e. the average of all available models) and for 1°C of global warming (IPCC 2013)

changes than on temperature changes. Figure 4.4 shows these changes for the average model and a 1°C change in global temperature. One can see a large-scale pattern, with increased precipitation in high latitudes and in the tropics, and a decrease in precipitation in mid-latitudes, and especially in Central America, the Mediterranean region, South Africa, and Australia.

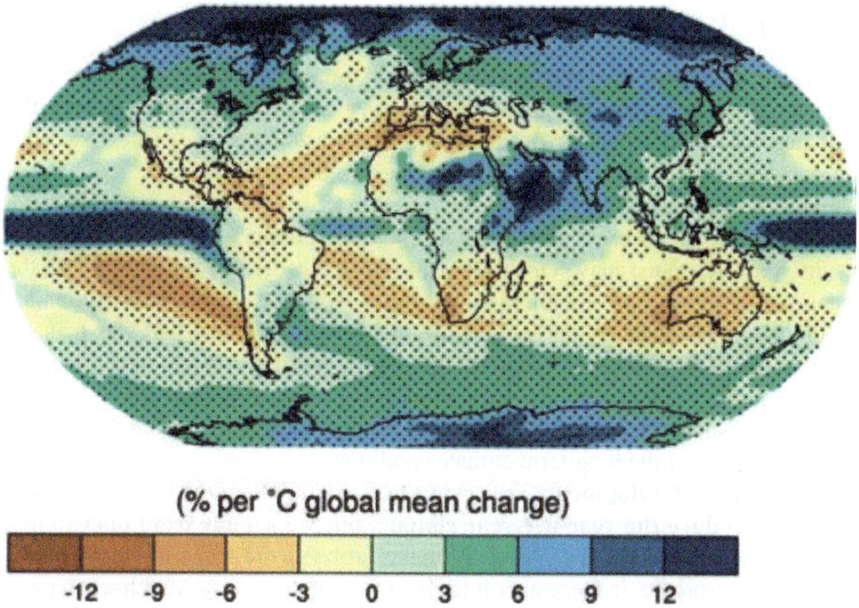


**Fig. 4.4** Precipitation changes (in mm/day per degree C of global temperature change) for the average model. Stippled regions are those where the mean signal is larger than the 95 percentile of the model dispersion, suggesting a strong signal (IPCC 2013)

Climate change will also cause a rise in global sea level. This rise is caused by two processes: (i) the increase in sea water temperature, which leads to water dilatation and increased sea volumes; (ii) the melting of continental icecaps in mountain regions and at high-latitudes (especially the Greenland ice sheet). The rate of sea level rise is extremely uncertain, because of our poor understanding of the mechanisms at play – especially regarding the melting of continental icecaps. IPCC (2013) estimates lie between 25 and 100 cm of sea level rise by 2100. Some assessments have been published with worst case scenarios higher than 150 cm, but there is a low confidence in these pessimistic results.

### *4.2.2 Forecasting Natural Variability*

**Climate variables show the overlap of (i) a long term trend due to human emissions of GHG and (ii) a natural variability due to the internal dynamics of the climate systems and non-human forcing (e.g., volcanic eruption, changes in solar influx).** This is why some decades show a quicker increase in temperature than others. In the 1990s, it is likely that natural variability enhanced the trend due to GHG emissions, making the temperature change extremely rapid. In the 2000s, on the other hand, natural variability and the human-caused trend are likely to have been in opposite direction, creating a plateau in global average temperature.

**GCMs reproduce the average climate and climate variability (e.g., how each year can be different to the next), but they are not designed to reproduce the year-per-year or decade-per-decade natural variability.** Some years or decades are warmer than others, and the models are supposed to reproduce the frequency of warmer vs. closer years (or decades), but not the occurrence of a warm or hot year (decade). The same model run twice with the same parameters will give different results; the difference between the two runs will be the uncertainty due to the chaotic nature of natural variability. As a result, it would be meaningless to compare the climate reproduced by the models in 2013 with the climate that was actually observed in 2013, or even to compare 10-year long periods. What makes sense is to compare the frequency of hot years over 30-year periods – e.g., the 1970–2000 period. Over such long periods, natural variability is averaged out, and the average is a fair estimate of the long-term climate conditions.

**Some recent developments are trying to improve the model to make them able to reproduce the year-per-year climate, but we are far from operational climate forecasts.** Models are able to make probabilistic forecasts for the next season (e.g., to predict in December if the next January-February-March season will be particularly cold, warm, wet or dry). The ability to do so is larger in tropical areas and for some large-scale atmospheric patterns. For instance, models can anticipate the El Niño – La Niña patterns, which have large impacts on many countries (e.g., on forest fires in Indonesia, drought in Latin America, and fisheries in the Pacific). In extra-tropical regions and especially Europe, the skills of these predictions are inexistent.

**Going beyond the season and forecasting natural variability over the next two decades in order to implement adaptation plans would be extremely useful but is out of reach.** Climate change is relatively limited at these time scales, and climate variability is such that the climate signal is not dominant. Consequently, climate models are incapable of predicting changes in the near future. It is therefore essential not to over-interpret the results of these models and not to use their output as forecasts, without taking into account natural variability. The inability of models to predict climate changes in the next two decades could change if work on the 10-year forecast – a focus of research today – progresses. This would nevertheless require considerable strides in numeric modeling and better knowledge of ocean conditions that determine climate change on these time scales. Improved knowledge requires more developed measurement networks in oceans worldwide.

## 4.3 "Downscaling" Global Climate Scenarios to Extreme Event Scenarios

To assess climate change impacts on extreme events, one usually starts from one or several socio-economic scenarios and uses one or several GCMs to create "climate scenarios", i.e. low-resolution global simulations of climate change. Their resolution is between 50 and 300 km.

This low spatial resolution is sometimes good enough to carry out impact analyses, e.g. to assess the impact on agriculture or forestry when micro-climates do not play a crucial role. Quite often, however, climate impacts will strongly depend on the precise local change in climate and a higher resolution is needed. In cities, local climate is particularly important because of their micro-climates including the urban heat island (UHI) effect (Oke 1987). UHI refers to temperatures that are often warmer downtown in comparison with the outskirts of the city, only a few kilometers away. The temperature differences can reach up to 10°C for large urban agglomerations and can strongly amplify heat stress, especially at night during heat waves, which in turn can lead to serious consequences in terms of public health. This was the case in 2003 when a strong heat wave occurred in Europe and caused more than 70,000 casualties with a higher percentage of victims in urban areas, for example in France.

In such a situation, it is necessary to downscale GCM output, to take into account high resolution processes, and produce climate scenarios at a resolution that is high enough to be used in impact models and analysis.

Some sort of downscaling is also needed when the phenomena that one wants to consider are too small to be adequately reproduced by GCMs. Examples of such phenomena are tropical cyclones or heavy precipitation that have a spatial scale of the order of a few kilometers. In the case of precipitation, the problem is complicated by the fact that distant precipitation changes can influence flood risks, if they occur upstream. This type of spatial connection may make it necessary to carry out the climate downscaling over a larger area, increasing the cost of doing so.

There are two ways of downscaling: using statistical methods or physical models.

### *4.3.1 Statistical Methods*

The first method uses statistical relationships, calibrated on historical data, to relate large-scale drivers – which GCMs can reproduce – to local phenomena – which GCMs cannot reproduce. Even though our knowledge of the laws of physics helps in selecting potential predictors, this method is not directly based on physical laws.

A classical example of this method is the use of statistical models for hurricane activity. Indeed, most climate models cannot reproduce realistic tropical cyclones and hurricanes, because of the very low resolutions of the mechanisms that need

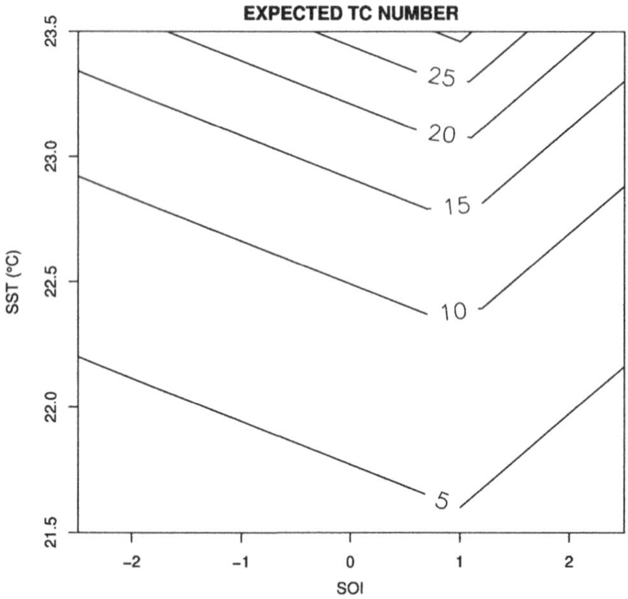


**Fig. 4.5** The number of tropical cyclones that should be expected in the North Atlantic, as a function of two large-scale climate parameters: the sea surface temperature in the North Atlantic, and the Southern Oscillation Index (a proxy for El Niño) (Source: Mestre and Hallegatte 2009)

to be simulated to do so. But climate models can reproduce well the large-scale characteristics of the climate in hurricane regions, such as the average sea surface temperature or the phase of large-scale climate oscillation like El Niño. Since these large-scale parameters are well correlated with hurricane activity, one can use historical data to assess the statistical relationship between large-scale climate parameters and hurricane activity, and use this relationship to estimate how hurricane activity will change in a different climate.

Such a model was applied by Elsner and Jagger (2006), who estimated the return level of extreme hurricane winds on the U.S. coastline, as a function of global climate indices like ENSO and NAO, which can be represented by GCMs. In Mestre and Hallegatte (2009), the link between the number of hurricanes in a year in the North Atlantic is estimated as a function of two large-scale parameters: the sea surface temperature in the North Atlantic, and the Southern Oscillation Index (a proxy for El Niño). Figure 4.5 shows how the expected number of tropical cyclones will change depending on these parameters. So even if a climate model is unable to reproduce directly tropical cyclones, the change in their frequency can be estimated by looking at how the sea temperature and the SOI parameters are expected to change, and the relationship of these two parameters with the number of tropical cyclones in the past.

Such statistical methods are computationally efficient and have often a good representation and ability to predict the current climate. Statistical models, however, have two main drawbacks: first, they need long series of reliable data; second, even with a sufficiently large data range, it is difficult to know whether a statistical relationship will remain valid in a future climate. A statistical relationship between large-scale climate indices and small-scale variables, indeed, can be different in a different climate. For instance, the correlation between sea surface temperature and hurricane intensity is very strong in the present climate (see, e.g., Emanuel 2005), but it does not mean that if the global climate warms by 2°C, hurricane intensity will automatically increase: the effect of a local or temporary perturbation may be different from the effect of a global or permanent change.

Also, if UHI is modeled with a statistical relationship calibrated on historical situations with large-scale temperatures that lie between 10 and 30°C, it is impossible to tell if the relationship remains valid when large-scale temperature exceeds 30°C. This situation will, however, arise when using this relationship to assess climate change impacts. Even more problematic, city infrastructure will change in the future: new buildings will be built, new neighborhoods will be developed, new parks will be introduced, and air-conditioning equipment will be installed. These new developments may modify the statistical link between large-scale climate indices and small-scale conditions. Statistical relationships, therefore, must be used with care, if possible in situations where conditions will not change too much.

### *4.3.2 Physical Models*

To avoid the problem of validity of historical relationships, one may use physical models, which are based on physical laws. Physical models are of particular interest when investigating extreme patterns and variability changes. Of course, physical models often require calibration, so that the distinction between physical models and statistical models is sometimes fuzzy.

These physical models can be Regional Climate Models (RCM) that take as input a large-scale forcing produced by GCM or specific models like hurricane models. Several large-scale international projects such as PRUDENCE, ENSEMBLES and then CORDEX, are based on this approach: GCMs are used to produce scenarios of global climate change at low resolution (between 50 and 200 km), and several RCM are then used to downscale these findings and produce climate change scenarios over Europe, at a resolution that allows impact analysis (at 10–50 km). Impacts can then be analyzed for different sectors (agriculture, water management, etc.).

To account for the specifics of urban areas, a 50 or even 25 km resolution is not enough and a better resolution is required to reproduce the UHI effect. To do so, specific urban models have been developed, with very high resolution and specific sub-modules to take into account characteristics of urban land cover. For instance, the Town Energy Balance (TEB) (Masson 2000) is a model that reproduces the energy fluxes in urban environments and that can be included in the high-resolution atmospheric model Meso-NH (with resolution of up to 250 m) that is, therefore, able to represent all aspects of the urban meteorology, including all kinds of land covers: natural soils, vegetation, water, and built-up areas. Numerous simulations can then be conducted for various urban environments and various conditions (e.g., Lemonsu et al. 2002; Pigeon et al. 2006, 2008), to assess how large-scale climate conditions and local urban interact to create urban micro-climates.

These developments have been used to get a better understanding of urban meteorology, but their application to climate change issues is still experimental. Only these models, however, would be able to predict the impact of higher global or regional temperature changes on street- or building-temperatures in cities, and to assess the effectiveness of adaptation measures like changes in building materials.

## 4.4 Consequences in Terms of Extremes

The large-scale changes in temperature, precipitations and other meteorological variables that models project suggest that all extreme events related to these variables (droughts, floods, heat waves, cold spells, etc.) will be affected. The IPCC conclusions are the following, for a few categories of extreme events:

### *4.4.1 Heat Waves and Cold Spells*

**It is virtually certain that heat waves will become more frequent and more intense in the future.** In most climate models and in most emission scenarios, the summer of the 2003 heat waves, which provoked more than 70,000 deaths, becomes an "average" summer in Europe (see Fig. 4.6). This means that, in this region, every other summer will be warmer than the 2003 one. The 2003 summer can be used to assess the need for climate change adaptation.

Fortunately, climate change will not make all extremes increase in the future. Cold spells have very large negative consequences and are expected to decrease both in frequency and intensity in the future.

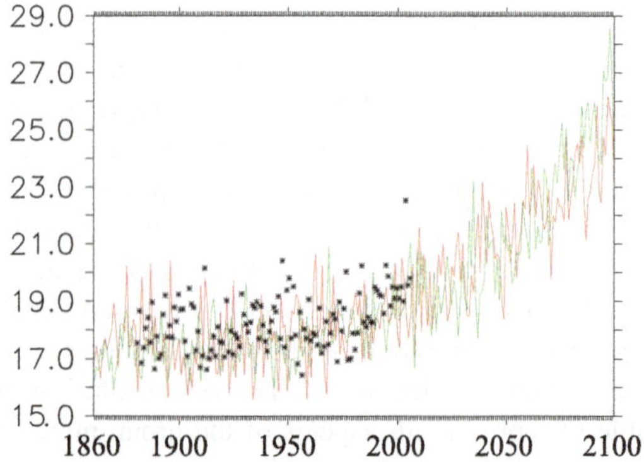


**Fig. 4.6** Average June-to-September temperature over France according to observations up to 2003, and from the IPSL (*green line*) and CNRM (*red line*) model simulations in the A2 emission scenario (Source: Jean-Louis Dufresne (IPSL))

### 4.4.2 *Droughts*

Water availability does not depend only on precipitations. An increase in water scarcity may arise from (i) changes in the annual amount of precipitations; (ii) from the deepening of the precipitation seasonal cycle, as more precipitation is projected in winter and less in summer; (iii) from the reduction of the buffering role of snow in mountain areas, since mountain winter precipitations are stocked under the form of a snow pack that melts slowly in spring and summer; (iv) from increased evaporation due to higher temperature.

**The evolution of droughts will vary depending on the location. It is likely, however, that many locations where droughts already are an issue will see the situation worsen.** Examples of such locations are the Mediterranean basin, the South-eastern Europe, South Africa, Australia, Central America. Some regions that currently experience regular water-stress are however expected to see an improvement in water resources, like East Africa.

### 4.4.3 *Storms and High Winds*

With higher temperatures, all atmospheric circulations will be modified, influencing winds and storms everywhere. To assess how wind hazards will change in the future, however, it is useful to distinguish between tropical storms and extra-tropical storms:

#### 4.4.3.1 Tropical Storms

Tropical storms are present in tropical regions and are particularly destructive. The strongest of them are referred to as hurricanes in the North Atlantic and typhoons in the Pacific.

After the unprecedented 2004 and 2005 hurricane seasons in the North Atlantic, some have argued that climate change has started to significantly modify hurricane characteristics: Webster et al. (2005) observed that hurricanes in the strongest categories (4 and 5) have almost doubled in number and in proportion in 30 years; over the last 75 years, Emanuel (2005) detected in the North Atlantic and western North Pacific basins a strong increase in the power-dissipation index (PDI), which is a proxy of the destructiveness of hurricanes. The debate on the significance and persistence of these trends, however, has yet to be resolved (Landsea 2005; Landsea et al. 2006).

The influence of climate change on hurricanes in the North Atlantic basin is a much debated topic and different approaches have predicted different futures. In observation analyses, the surface temperature of the North Atlantic is driving hurricane activity (e.g., Mestre and Hallegatte 2009). Vecchi and Soden (2007) suggest however that the driver is the temperature difference between the tropical storm area and the rest of the tropics, questioning the relationship between global temperature and hurricane activity.

Emanuel et al. (2008) use a hurricane model that takes as input large-scale conditions (wind patterns, thermodynamic conditions, etc.), and provides statistics on hurricane tracks and intensity. Analyzing seven different climate models, they conclude that climate change could significantly modify hurricane risks, although the sign and magnitude of the changes vary from basin to basin and from model to model, reflecting large differences in model projections at the regional scale.

In Global Circulation Models (GCMs), hurricanes seem to be only marginally affected by higher temperatures (e.g., Chauvin et al. 2006; Murakami and Wang 2010). But the low resolution of these models makes their findings on hurricanes questionable. Downscaling approaches have logically been proposed: using a Regional Climate Model (RCM), Knutson et al. (2013) predict a significant reduction in tropical storm frequency (by 20–30 %). However, lifetime maximum hurricane intensity increases in their analysis, by 4–6 %. There is an increase of 40–90 % in the frequency of very intense (categories 4 and 5) hurricanes. Also, they find rainfall rates that increase by +20 to +30 % in the hurricane's inner core (and a smaller increase (~10 %) at 200 km from the center of the storm).

The rise in average sea level will however make hurricanes more dangerous. Indeed, hurricanes create "storm surges", i.e. temporary rises in sea level due to lower air pressure and wind effects. The combination of a higher average sea level with temporary rise due to storm can have dire impacts through coastal floods. See an analysis of these effects on New Orleans in Chap. 6.

**There is a large uncertainty on the future of hurricanes and tropical cyclones. A best guess today is that the number of tropical storms may decrease in the future, but that the number of the most intense storms may increase, especially in the North Atlantic. An analysis of the impact of such change in the US is provided in Sect. 5.2.**

#### 4.4.3.2 Extra-Tropical Storms

Extra-tropical storms cause significant losses in mid to high-latitude regions. Examples of such storms are those that struck Europe in December 1999, Lothar and Martin. There is no agreement on the evolution of extra-tropical storms in the future among climate models, with some models projecting a small increase in maximum wind speeds, while others project a small decrease. Models also suggest the possibility of a shift toward higher latitudes of the mean trajectory of these storms. Generally speaking, however, little change is expected for extra-tropical storms. The main impact of climate change on the impact of extra-tropical storms is likely to come from sea level rise, which will make storm surge more destructive (see Sect. 5.3 for an illustration on Copenhagen).

### 4.4.4 River Floods

As precipitations change, river run-off will change too. This change may have large consequences on flood risks. The average run-off is a weak proxy for changes in river flood, because a decrease in average precipitation does not imply that extreme precipitations cannot increase. In some regions, however, flood risks will decrease as precipitations diminish. But in other regions, flood risks will increase with potentially large economic consequences. Figure 4.7 shows the change in the frequency of what is currently a 100-year flood (i.e. a flood that has a 1 % probability of occurring each year), over metropolitan France. It highlights the heterogeneity, with the probability of the event doubling – or being halved – depending on the

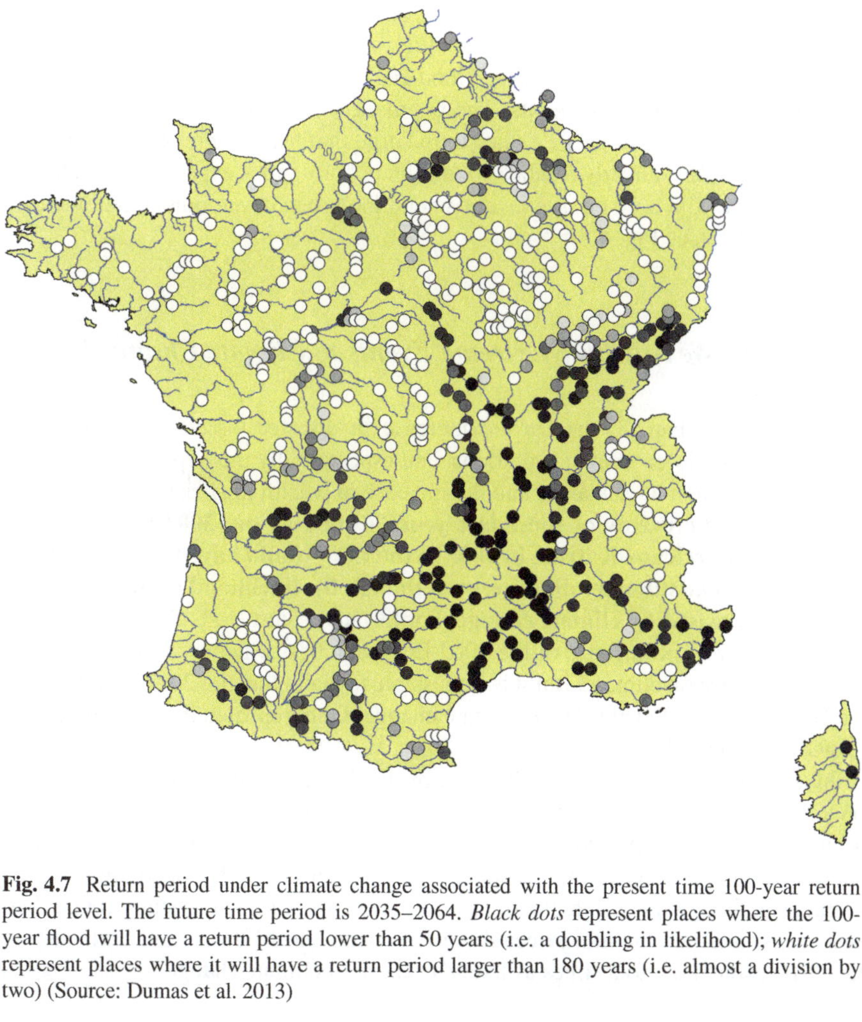

**Fig. 4.7** Return period under climate change associated with the present time 100-year return period level. The future time period is 2035–2064. *Black dots* represent places where the 100-year flood will have a return period lower than 50 years (i.e. a doubling in likelihood); *white dots* represent places where it will have a return period larger than 180 years (i.e. almost a division by two) (Source: Dumas et al. 2013)

location. In particular, climate change is likely to have a positive impact on the most intense precipitation in many places (Min et al. 2011), with serious consequences on flash floods and river floods.

### *4.4.5 Coastal Floods*

Climate change will cause a rise in sea level, which will increase the flood risk from storm surges – i.e. high-frequency high water levels due to low pressure and high winds created by a storm. In fact, it is likely that the largest losses from sea level rise will not be due to a rise in mean sea level, but from its short-duration extremes. These risks are already large, as illustrated by the disastrous consequences of the tropical cyclone Sidr in Bangladesh in 2007, or the destructions caused by hurricane Katrina in New Orleans in 2005. Large cities are particularly vulnerable: Hanson et al. (2011) investigated the coastal flood risk in all world cities with currently more than one million inhabitants. The population of these cities that is exposed to the 100-year storm surge – i.e. the storm surge that has an annual probability of occurrence of 1 % – already reaches 40 million people, and is expected to reach 140 million people with a 50 cm sea level rise and the urban development that one can expect by 2070 (see Fig. 4.8).

### *4.4.6 Can We Attribute Extreme Events to Climate Change?*

As stressed in Chap. 3, the current increase in disaster losses can be entirely explained by socio-economic evolutions, and especially by the population increase in coastal and other at-risk areas and the increase in wealth in at-risk areas.

**The dramatic events of these last years – such as the 2003 heat wave in Europe and hurricane Katrina in 2005 in the US – would have been possible in a climate with no human influence, even though the likelihood of some of them has been increased by climate change.** For instance, Stott et al. (2004) claim that the probability of occurrence of the 2003 European heat wave was doubled by the human influence on climate. But it is impossible to claim that the 2003 heat wave or the UK floods in 2000 are *due to* climate change. Only repeated heat waves can be related to climate change.

In a 2013 special issue of the Bulletin of the American Meteorological Society (Peterson et al. 2013), 19 analyses were performed on 12 extreme events that occurred in 2012. Approximately half the analyses found some evidence that anthropogenically caused climate change was a contributing factor to the extreme event, even though natural fluctuations also played a key role. The study found that heatwaves, like the one that affected the American and Midwest and northeast in July 2012, were now four times as likely, because of climate change.

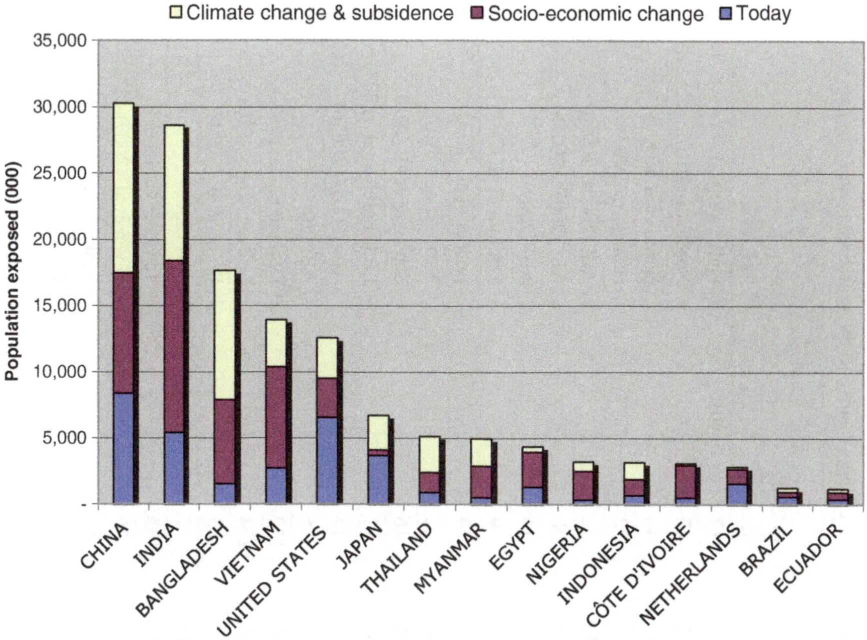


**Fig. 4.8** Increase between 2005 ("today") and the 2070s in population exposed to the 100-year coastal floods in coastal cities (of more than one million inhabitants in 2005). The figure shows the role of climate change (assumed to lead to 50 cm of sea level rise and a 10 % increase in storm frequency) and subsidence, and the role of socio-economic change (from an OECD scenario). At the global scale, climate change and subsidence are responsible for one third of the total increase, but this ratio varies depending on countries (Source: Hanson et al. 2011)

**Generally, climate change should be considered as a change in the frequency of weather events: no single event can be considered as a consequence of climate change, only the repetition of some types of event can be attributed to climate change.**

**In some cases, however, recent hazards with large consequences can be considered as an example of what will occur more frequently in the future.** If one cannot say that the 2003 heat is due to climate change, one can say that such an event will become more frequent and that adaptation actions are required to increase our resilience to high temperatures.

## 4.5 How Would These Changes in Hazard Translate into Changes in Losses?

A change in hazard does not translate simply into changes in disaster frequency and intensity and economic losses. Table 4.1, extracted from Bouwer (2013) and IPCC (2012), summarizes assessments made for tropical cyclones, extra-tropical storms, and river floods in various regions.

**Table 4.1** Estimated change in extreme weather losses in 2040 due to climate change and exposure change, relative to the year 2000 from 21 impact studies

| A. Climate change | | | | | | | |
|---|---|---|---|---|---|---|---|
| | | | Estimated loss change [%] in 2040 | | | | |
| No. | Hazard type | Region | Min | Max | Mean | Median | Reference |
| 1 | Tropical cyclone | Global | 58 | 1,365 | 417 | 30 | Pielke (2007) |
| 2 | Tropical cyclone | USA | 12 | 92 | 47 | | Nordhaus (2010) |
| 3 | Tropical cyclone | Global, low lat. | 23 | 130 | 46 | | Narita et al. (2009) |
| 4 | Tropical cyclone | USA | – | – | 22 | | Hallegatte (2007) |
| 5 | Tropical cyclone | USA, Caribbean | 19 | 46 | 32 | | ABI (2005a, 2005b) |
| 6 | Tropical cyclone | Japan | 20 | 45 | 30 | | ABI (2005a, 2005b) |
| 7 | Tropical cyclone | China | 9 | 19 | 14 | | ABI (2009) |
| 8 | Tropical cyclone | USA | – | – | 9 | | Schmidt et al. (2009) |
| 9 | Tropical cyclone | USA | −27 | 36 | 14 | | Bender et al. (2010) |
| 10 | Extra-tropical cyclone | Global, high lat. | −11 | 62 | 22 | 15 | Narita et al. (2010) |
| 11 | Extra-tropical cyclone | Europe | 6 | 25 | 16 | | Schwierz et al. (2010) |
| 12 | Extra-tropical cyclone | UK, Germany | −6 | 32 | 11 | | Leckebusch et al. (2007) |
| 13 | Extra-tropical cyclone | Europe | – | – | 14 | | ABI (2005a, 2005b) |
| 14 | Extra-tropical cyclone | UK | −33 | 67 | 15 | | ABI (2009) |
| 15 | Extra-tropical cyclone | Netherlands | 80 | 160 | 120 | | Dorland et al. (1999) |
| 16 | River flooding | Netherlands | 46 | 201 | 124 | 83 | Bouwer et al. (2010) |
| 17 | River flooding | Europe | – | – | 83 | | Feyen et al. (2009) |
| 18 | River flooding | UK | 3 | 11 | 7 | | ABI (2009) |
| 19 | River flooding | Rhine Basin | 57 | 213 | 135 | | Te Linde et al. (2010) |
| 20 | River flooding | Spain (Madrid) | – | – | 36 | | Feyen et al. (2009) |
| 21 | River flooding | Australia | 67 | 514 | 361 | | Schreider et al. (2000) |
| 22 | Local flooding | Netherlands | 16 | 70 | 47 | | Hoes (2007) |

B. Exposure change

| No. | Hazard type | Region | Estimated loss change [%] in 2040 | | | Reference |
|---|---|---|---|---|---|---|
| | | | Min | Max | Mean | |
| 1 | Tropical cyclone | Atlantic | 164 | 545 | 355 | Pielke (2007) |
| 8 | Tropical cyclone | USA | – | – | 240 | Schmidt et al. (2009) |
| 15 | Extra-tropical cyclone | Netherlands | 12 | 93 | 50 | Dorland et al. (1999) |
| 16 | River flooding | Netherlands | 35 | 172 | 104 | Bouwer et al. (2010) |
| 19 | River flooding | Rhine Basin | 10 | 36 | 23 | Te Linde et al. (2010) |

Source: Bouwer (2013)

**The table shows the uncertainty in the assessment of the effect of climate change on disaster losses, and the potential for large changes in losses.** It also shows that for most studies, the impact of socio-economic factors dominates the effect on climate change on future disaster losses. These results need to be considered with care, however.

**First, an increase in disaster losses due to economic growth is compensated by the larger coping and recovering capacity that higher incomes bring.** As a result, an increase in losses due to economic growth does not mean an increase in vulnerability, while an increase due to climate change corresponds most of the time to an increase in vulnerability.

**Second, most of these analyses are unclear on the assumptions made on adaptation. A recent study of 136 coastal cities** (Hallegatte et al. 2013) **shows that with no adaptation at all, losses can soar extremely rapidly, but such a scenario is not realistic.** In the 136 cities, annual flood losses are found to increase from $6 billion per year today to more than one trillion US dollars per year if nothing is done. This is because current defenses rapidly become ill-adapted to new climate conditions. With higher average sea level and no change in defenses, some places are flooded very frequently, and losses become unacceptable: in some places, the absence of defenses upgrading would mean that floods would occur every few years.

This scenario is clearly unrealistic: after repeated floods, an upgrade of defenses would be implemented. And the same study finds that $50 billion per year of adaptation expenditure could cancel the increase of flood losses; a large investment, but a very cost-effective one since it would cancel most of one trillion US dollar in annual disaster losses. **The large increase in losses in scenarios with no adaptation simply demonstrates that not adapting is simply not an option, and that adaptation will take place in spite of its cost.**

## 4.6 Conclusion

**Climate change will not increase all extreme events everywhere. But climate change is very likely to increase some extreme events and hazards, in some regions. For hazards for which detailed analyses are available, it is often the most intense events that are found to increase in response to climate change.**

**There is little literature on how climate change mitigation – i.e. reduction in emissions of GHG – can affect the frequency and intensity of extreme events over different timescales.** In other terms, it is unclear how efficient climate change mitigation will be in reducing future losses from extreme events. One important study found that limiting global warming to 2°C (compared with the pre-industrial global temperature in 1850) could halve climate change damages (Arnell et al. 2013). This reduction in climate change damages is partly due to a reduction in disaster losses.

**The impact of climate change mitigation on extreme events will also depend on the type of extreme events**: sea level rise is a very slow process that cannot be reversed or even slowed down rapidly, and climate change mitigation takes a long time to reduce risks from coastal floods. On the other hand, temperature reacts more rapidly to reduced emissions, and climate change mitigation can reduce heat wave risks over a few decades.

**Published estimates suggest that disaster losses will increase with climate change.** But different adaptation scenarios would lead to very different results. The next chapter focuses on how to assess the impact of climate change on disaster losses taking into account the adaptation and risk management policies that can be implemented.

## References


ABI (2005a) Financial risks of climate change: summary report. Association of British Insurers, London

ABI (2005b) Financial risks of climate change: technical annexes. Association of British Insurers, London

ABI (2009) The financial risk of climate change. Research paper no. 19, Association of British Insurers, London

Arnell NW, Lowe JA, Brown S, Gosling SN, Gottschalk P, Hinkel J, Lloyd-Hughes B et al (2013) A global assessment of the effects of climate policy on the impacts of climate change. Nat Clim Chang 3(5):512–519

Bender MA, Knutson TR, Tuleya RE, Sirutis JJ, Vecchi GA, Garner ST, Held IM (2010) Modeled impact of anthropogenic warming on the frequency of intense Atlantic hurricanes. Science 327:454–458

Bouwer LM (2013) Projections of future extreme weather losses under changes in climate and exposure. Risk Anal 33(5):915–930

Bouwer LM, Bubeck P, Aerts JCJH (2010) Changes in future flood risk due to climate and development in a Dutch polder area. Glob Environ Chang 20:463–471

Chauvin F, Royer J-F, Déqué M (2006) Response of hurricane-type vortices to global warming as simulated by ARPEGE-Climat at high resolution. Clim Dyn 27:377–399

Dorland C, Tol RSJ, Palutikof JP (1999) Vulnerability of the Netherlands and Northwest Europe to storm damage under climate change. Clim Chang 43:513–535

Dumas P, Hallegatte S, Quintana-Seguí P, Martin E (2013) The influence of climate change on flood risks in France-first estimates and uncertainty analysis. Nat Hazards Earth Syst Sci 13:809–821

Elsner JB, Jagger TH (2006) Prediction models for annual U.S. hurricane counts. J Clim 19: 2935–2952

Emanuel K (2005) Increasing destructiveness of tropical cyclones over the past 30 years. Nature 436:686–688

Emanuel K, Sundararajan R, Williams J (2008) Hurricanes and global warming: results from downscaling IPCC AR4 simulations. Bull Am Meteorol Soc 89(3):347–367

Feyen L, Barredo JI, Dankers R (2009) Implications of global warming and urban landuse change on flooding in Europe. In: Feyen J, Shannon K, Neville M (eds) Water and urban development paradigms. Taylor and Francis, London, pp 217–225

Hallegatte S (2007) The use of synthetic hurricane tracks in risk analysis and climate change damage assessment. J Appl Meteorol Climatol 46(11):1956–1966

Hallegatte S, Green C, Nicholls R, Corfee-Morlot J (2013) Flood losses in major coastal cities. Nat Clim Chang 3(9):802–806

Hanson S, Nicholls R, Ranger N, Hallegatte S, Corfee-Morlot J, Herweijer C, Chateau J (2011) A global ranking of port cities with high exposure to climate extremes. Clim Chang 104(1): 89–111

Hoes OAC (2007) Aanpak wateroverlast in polders op basis van risicobeheer. PhD thesis, Delft University of Technology, Delft (in Dutch)

IPCC (2012) Managing the risks of extreme events and disasters to advance climate change adaptation. A special report of Working Groups I and II of the Intergovernmental Panel on Climate Change [Field CB, Barros V, Stocker TF, Qin D, Dokken DJ, Ebi KL, Mastrandrea MD, Mach KJ, Plattner G-K, Allen SK, Tignor M, Midgley PM (eds)]. Cambridge University Press, Cambridge/New York, 582 pp

IPCC (2013) Climate change 2013: the physical science basis. Contribution of Working Group I to the fifth assessment report of the Intergovernmental Panel on Climate Change [Stocker TF, Qin D. Plattner G-K, Tignor M, Allen SK, Boschung J, Nauels A, Xia Y, Bex V, Midgley PM (eds)]. Cambridge University Press, Cambridge/New York, 1535 pp

Knutson TR et al (2013) Dynamical downscaling projections of twenty-first century Atlantic hurricane activity: CMIP3 and CMIP5 model-based scenarios. J Clim 26:6591–6617

Landsea CW (2005) Meteorology: hurricanes and global warming. Nature 438:E11–E13

Landsea CW, Harper BA, Hoarau K, Knaff JA (2006) Can we detect trends in extreme tropical cyclones? Science 313(5786):452–454

Leckebusch GC, Ulbrich U, Fröhlich L, Pinto JG (2007) Property loss potentials for European midlatitude storms in a changing climate. Geophys Res Lett 34:L05703

Lemonsu A, Masson V (2002) Simulation of a summer urban breeze over Paris. Bound-Lay Meteorol 104(3):463–490

Masson V (2000) A physically-based scheme for the urban energy budget in atmospheric models. Bound-Layer Meteorol 94(3):357–397

Mestre O, Hallegatte S (2009) Predictors of tropical cyclone numbers and extreme hurricane intensities over the North Atlantic using generalized additive and linear models. J Climate 22(3):633–648

Min S-K, Zhang X, Zwiers F, Hegerl G (2011) Human contribution to more-intense precipitation extremes. Nature 470:378–381

Murakami H, Wang B (2010) Future change of North Atlantic tropical cyclone tracks: projection by a 20-km-mesh global atmospheric model. J Clim 23:2699–2721

Nakicenovic N, Swart R (2000) Special report on emissions scenarios: a special report of Working Group III of the Intergovernmental Panel on Climate Change. Cambridge University Press, Cambridge, 599 pp

Narita D, Tol RSJ, Anthoff D (2009) Damage costs of climate change through intensification of tropical cyclone activities: an application of FUND. Clim Res 39:87–97

Narita D, Tol RSJ, Anthoff D (2010) Economic costs of extratropical storms under climate change: an application of FUND. J Environ Plan Manag 53:371–384

Nordhaus WD (2010) The economics of hurricanes and implications of global warming. Clim Chang Econ 1:1–20

Oke TR (1987) Boundary layer climates. Cambridge University Press, Cambridge

Peterson TC, Hoerling MP, Stott PA, Herring S (eds) (2013) Explaining extreme events of 2012 from a climate perspective. Bull Am Meteorol Soc 94(9):S1–S74

Pigeon G, Lemonsu A, Long N, Barrié J, Masson V, Durand P (2006) Urban thermodynamic island in a coastal city analysed from an optimized surface network. Bound-Lay Meteorol 120(2): 315–351

Pigeon G, Moscicki MA, Voogt JA, Masson V (2008) Simulation of fall and winter surface energy balance over a dense urban area using the TEB scheme. Meteorol Atmos Phys 102(3–4): 159–171

Pielke RA Jr (2007) Future economic damage from tropical cyclones: sensitivities to societal and climate changes. Phil Trans R Soc A 365:2717–2729

Schmidt S, Kemfert C, Höppe P (2009) Tropical cyclone losses in the USA and the impact of climate change: a trend analysis based on data from a new approach to adjusting storm losses. Environ Impact Assess Rev 29:359–369

Schreider SY, Smith DI, Jakeman AJ (2000) Climate change impacts on urban flooding. Clim Chang 47:91–115

Schwierz C, Köllner-Heck P, Zenklusen Mutter E, Bresch DN, Vidale PL, Wild M, Schär C (2010) Modelling European winter wind storm losses in current and future climate. Clim Chang 101:485–514

Stott PA, Stone DA, Allen MR (2004) Human contribution to the European heatwave of 2003. Nature 432(7017):610–614

Te Linde AH, Aerts JCJH, Bakker AMR, Kwadijk CJC (2010) Simulating low-probability peak discharges for the Rhine basin using resampled climate modeling data. Water Resour Res 46(3):W03512

van Vuuren DP, Edmonds J, Kainuma M, Riahi K, Thomson A, Hibbard K, Hurtt GC et al (2011) The representative concentration pathways: an overview. Clim Chang 109(1–2):5–31

Vecchi GA, Soden BJ (2007) Effect of remote sea surface temperature change on tropical cyclone potential intensity. Nature 450(7172):1066-U9

Webster PJ, Holland GJ, Curry JA, Chang H-R (2005) Changes in tropical cyclone number, duration, and intensity in a warming environment. Science 309:1844–1846

# Chapter 5
# Climate Change Impact on Natural Disaster Losses

**Abstract** This chapter proposes a methodology to assess the impact of climate change on natural hazards and disaster losses, and applies it to three local case studies: hurricanes in the United States; storm surges in Copenhagen; and heavy precipitation in Mumbai. These three case studies are chosen to illustrate different aspects of the questions and the different tools that can be mobilized to provide inputs and information to decision-makers. The chapter concludes that stronger hazards due to climate change do not need to translate into more or larger disasters, as proactive risk analysis and management approaches can be used to cope with this trend.

**Keywords** Climate change • Natural hazards • Disaster losses • Assessment methodology

**It is difficult to assess how the effect of climate change on natural hazards can affect natural disaster losses. Doing so requires carrying out detailed analysis at local scale, In the face of significant methodological problems.** Local analyses indeed require a lot of data and modeling capacity that may not be available; for instance, most developed world river basins have been modeled and these models can be used to investigate climate change impacts; but for the developing world, there is no such toolbox of models, and even basic data (e.g., on topography or economic exposure) is often inexistent. In spite of these problems, local analyses are necessary. First, as shown in the previous chapter, the climate change effect on extreme events and hazards is extremely heterogeneous and a local analysis is needed to account for this heterogeneity. Second, as shown in Chap. 2, the impact of a physical hazard is highly dependent on local socio-economic characteristics, and translating the change in hazard into a change in economic and welfare losses requires a context-specific analysis.

**This chapter proposes a methodology to perform local-scale analyses of climate change impacts, and illustrates it in three case studies of hurricanes on the U.S. coast, on storm surges in Copenhagen, and on heavy precipitation on Mumbai.**

**These case studies consider the change in hazard and the local exposure, sensitivity and vulnerability. They also investigate the resilience of the affected**



S. Hallegatte, *Natural Disasters and Climate Change*,
DOI 10.1007/978-3-319-08933-1_5

**systems, by considering indirect impacts (and how they can be reduced). They show that losses can increase very rapidly with changes in hazard characteristics, and that the link between changes in hazards and changes in losses is highly nonlinear. They also look at adaptation options on the local scale, to assess the local adaptive capacity.** Very cost-effective options are identified in some cases – sometimes even at a negative cost – and these options (sometimes referred to as "no regret" options) are often linked to pre-existing sub-optimalities, i.e. the fact that we are not perfectly adapted to the current climate and the current hazards).

The three studies show that vulnerability depends on many factors that vary widely, such as extreme event frequency and intensity, local elevation and topography, building types and norms, but also the economic structure, the wealth level, and institutional arrangements (e.g., the existence and regulation of insurance markets). As a result of this complexity, estimating the vulnerability of any one city is a significant and time-consuming task, based on a large amount of local data and information.

## 5.1 Methodology for Local Assessment of Climate Change Impacts on Disaster Risks[1]

A first step in climate change impact assessment is to select (or develop) a global socio-economic baseline scenario, which determines how much climate change can be expected (see Fig. 5.1). Often, the global scenario also includes a vision of how different world regions will develop over time, as a function of demographic, economic and technologic trends and of political choices, including climate and mitigation policies (STEP 1). This information can be useful to inform the local analysis, for instance by providing boundary conditions to the analysis (e.g., global oil and carbon prices can be useful inputs to assess the costs and benefits of adaptation options that consume energy; the existence of an international system for post-disaster support also influences local vulnerability).

To allow for local impact analysis, it is also necessary to downscale the global socio-economic scenario to describe socio-economic conditions at the local scale (STEP 2). Creating such local scenarios leads to very specific issues. A key challenge is factoring in different views about the future. The future economy that will be impacted by climate change will differ from today's economy, and even small changes in economic development can make a difference in climate change impacts. For example, a reduction in poverty and an improvement in housing quality can reduce vulnerability to climate change. Impacts will therefore vary with levels and types of economic development as well as with levels and types of climate change.

[1] This section is based on Hallegatte et al. (2011a).

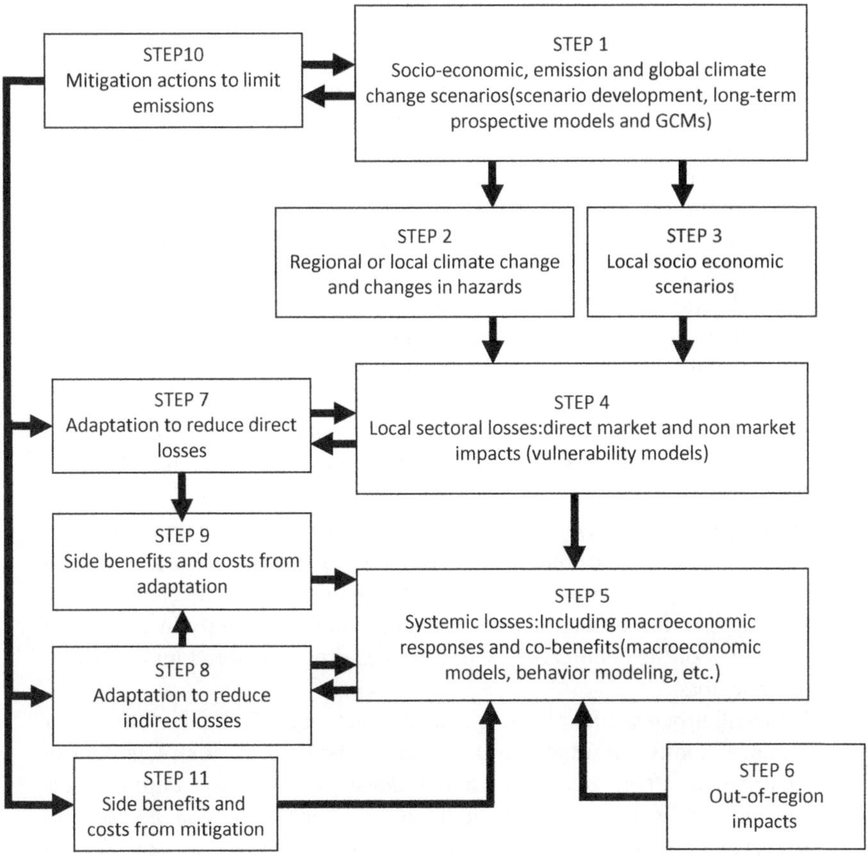


**Fig. 5.1** The different components necessary to assess climate change impacts at the local level

The specific case of scenarios for urban areas is detailed in Box 5.1, and shows that many additional assumptions are required to create local scenarios, introducing additional uncertainty compared with global scenarios.

**Box 5.1: Developing City-Scale Scenarios**

To develop long-term urban scenarios, it is possible to begin with global scenarios and layer in a number of "local perspectives" about key issues including:

- How will urbanization rates evolve? In particular, an important question is whether, with economic growth, developing countries will converge to

(continued)

**Box 5.1** (continued)

developed countries' urbanization rates, which are frequently higher than 80 %, or whether they will converge to a different end-point.

- How will additional urbanization in the country or region be distributed among cities? In particular, in the future, will city sizes become saturated at some level or will they keep growing? One possibility is that negative externalities (e.g. congestion and local pollution) will exceed the positive externalities of agglomeration (e.g. availability of workers, firm network externalities) and represent a strong obstacle to the development of such mega-cities. The fact that second-tier cities are currently growing more rapidly than the largest ones suggest that there is a saturation in city size.
- How will sub-regional population and economic growth differ within a country or a region? For example will coastal cities grow more rapidly than inland cities?
- How will economic development in cities differ from economic development in the rest of the country/region? Today, cities are wealthier than rural areas and they are growing more rapidly, but this trend could change in the future. Also, the economic structure will evolve differently in cities than in the rest of the country, e.g. with services growing more rapidly than in other sectors.
- How will urban spatial planning and architecture evolve in the future? For instance, additional urban development can be done through low-density suburban development or through high-density urbanization. Future streets can be narrower or wider than today; new materials can be used in streets and buildings, absorbing more or less solar energy; more parks can be introduced in cities. Depending on these urbanization characteristics, energy consumption, and exposure and vulnerability to heat waves and floods may differ significantly.
- How can infrastructure development keep pace with growing population? For instance, sewage and water treatment infrastructure are underdeveloped in many cities of the developing world (e.g. World Bank 2012). The development and the characteristics of future infrastructure will play a major role in city vulnerability to climate change. For instance vulnerability to increased flood incidence will depend on whether sewage and drainage systems are able to cope and on whether urban transport is based on roads and individual cars or on public transportation (e.g. subway).
- In coastal cities, vulnerability to coastal floods and storm surges will depend on human-induced subsidence that, in turn, depends in part on where the water consumed in the city is extracted. If ground water is pumped into the city, subsidence rates increase and amplify global sea level rise. Subsidence will depend on choices made by city planners.

(continued)

**Box 5.1** (continued)

For instance, the city of Shanghai has invested in reducing ground water pumping, thereby reducing human-induced subsidence.

While socio-economic scenario development is complex, their construction at local scale is both a necessary input and an essential part of any local climate impact and policy assessment and thus requires research attention. Moreover, the fact that future impacts depend on these factors makes it possible to use such scenarios to design policy levers to alter development pathways so as to reduce future climate change vulnerability and increase resilience.

Figure 5.2 reproduces one urbanization scenario developed in Viguie et al. (2013) for the Paris agglomeration using the NEDUM-2D model (Viguié and Hallegatte 2012). This scenario is a function of a global scenario produced by a global energy-economic model (IMACLIM-R, Rozenberg et al. 2010) and of hypotheses on local planning and transport policies. Such a spatial scenario has a clear influence on the exposure to many natural hazards, including floods and heat waves. Another approach to developing urban scenarios is provided in Solecki and Olivieri (2004).

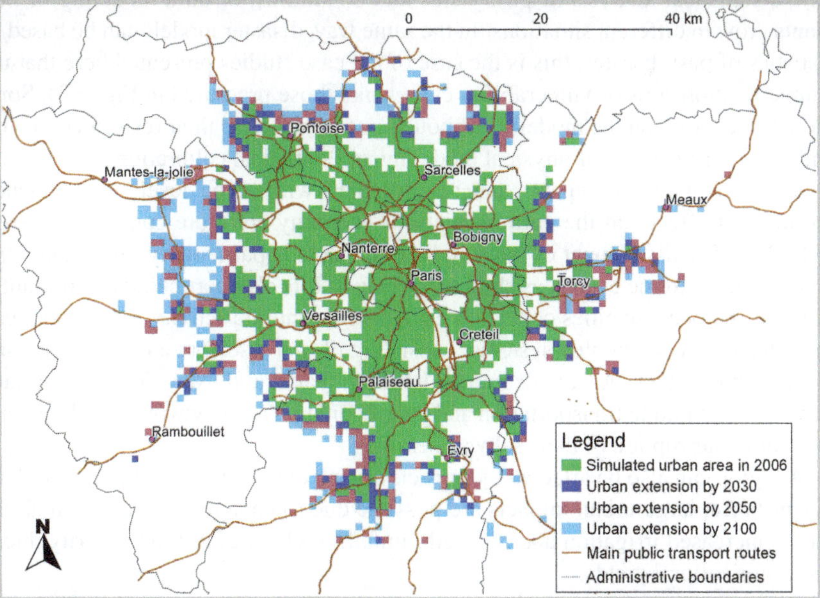


**Fig. 5.2** Extension of Paris agglomeration between 2010 and 2100, according to one scenario (Source: Viguie et al. 2013)

Future economies are not the only unknown components of the analysis as the future climate is also uncertain. Global emission and climate scenarios can be derived from global socio-economic scenarios but, similarly, climate scenarios are not at a sufficient resolution to carry out local impacts analysis. Downscaling of global or regional climate conditions must be done using statistical relationships or physical methods (see Chap. 4) (STEP 3).

From this downscaled information, direct sectoral economic impacts of climate change can be estimated in any region and sector; see examples in the case studies (STEP 4). These impacts arise partly from the change in mean climate, and partly from extreme events, hazards, and disasters. These sectoral and regional impacts are usually calculated using specific models, such as:

- Agriculture models that predict the yields of various production types as a function of environmental conditions (and farmers' actions).
- Energy models that predict energy production and demand as a function of climate conditions.
- Disaster loss models that predict direct (and/or indirect) disaster losses as a function of the physical characteristics of an event.

These models can be statistical models or physical models. For instance, a model predicting agricultural yields as a function of environmental conditions can be based on the statistical analysis of past yields and conditions, or on a modeling of how plants grow in different situations. In the same way, disaster models can be based on statistics of past disaster; this is the case of the case studies presented here that use simple relationships or vulnerability curves like those presented in Fig. 3.3). Some models are based on the modeling of housing structures and their resistance to wind and water, or on a mix of physical modeling and statistical calibration.

Even when direct impacts can be estimated with some level of confidence, the indirect effects on the entire society or economy are more complex to assess (STEP 5). The assessment of these indirect impacts is particularly important when considering extreme events, as shown in Chap. 2. It is also particularly important in urban areas, because cities concentrate so much economic activity in limited areas; they function as integrated systems, consisting of many sectors and infrastructures closely interlinked. As an example, London, Paris and New York are totally dependent on public transportation and their economic activity would be threatened by a long interruption of their subway service.

A region can also be affected by impacts arising outside its boundaries (STEP 6). For instance, large urban impacts are possible due to changes in water availability due to increased irrigation and reduced rainfall, or changes in food security due to lower agricultural yields.

Given the extent of uncertainties at each step of any assessment of impacts, and especially with respect to the global carbon cycle response to emission scenarios,[2] the local climate response, and local socioeconomic scenarios (including the

[2]This includes the important parameter of the climate sensitivity or the change in temperature that results from a doubling of carbon dioxide concentrations in the atmosphere.

influence of policy), it is essential not only to look at median (or "best guess") scenarios of change but also to look at the most extreme scenarios even if they have a small likelihood of occurring (Nicholls and Leatherman 1996; Fankhauser et al. 1999; Manning et al. 2004; Hallegatte 2009). Indeed, they could lead to significant losses and thus may justify specific anticipatory, preventive action (Jamet and Corfee-Morlot 2009; Weitzman 2009). Using an approach to bracket the uncertainty and look at the full range of plausible outcomes to examine the economics of city scale impacts helps to illustrate the link between development and climate mitigation, a cross-scale problem that is not well understood today.

Of course, all economic agents will respond to climate impacts by implementing adaptation strategies, therefore reducing damages. They will do so to reduce direct impacts (STEP 7), for instance through changes in urban plans or building norms to reduce flood-related damages to buildings. They will also do so to reduce indirect losses (STEP 8), for instance through the creation of an insurance scheme or a special fund to ease reconstruction after disasters or through the diversification of the local economy. These actions will have positive and negative co-benefits that need to be accounted for (STEP 9).

Their ability to cope with the new climate, however, is difficult to predict as it depends on the ability to detect a change in climatic conditions in due time, to develop technical or institutional responses to this change, and to implement these responses in an efficient way (Hallegatte et al. 2007). Past experience shows that detection failure, mal-adaptation, and over-reactions are common, and that adaptation can be delayed, or never happen (Yohe et al. 1995, 1996, 2011; Tol et al. 1998; Klein 2001). Chapters 6 and 7 discuss methodologies to maximize the potential from adaptation.

Finally, mitigation actions to reduce global climate change (i.e. to reduce emissions of greenhouse gases) can be accounted for (STEP 10), taking into account the possible co-benefits and co-costs (STEP 11) (Box 5.2).[3]

**Box 5.2: Using Impact Assessments to Inform Mitigation Decision?**
In an ideal world, adaptation and mitigation costs and benefits would be calculated at the global scale in a unique step through an optimization process (see a very comprehensive analysis of this coupled optimization in Lecocq and Shalizi 2007). However, as mentioned in Lecocq and Shalizi (2007) and Tol (2005), there are key differences in drivers or determinants of mitigation and

(continued)

[3]Of course there is clearly a difference between a decision on long-term climate change goals at the global scale and the actual delivery of some level of mitigation which when combined with other investment and development decisions will determine actual emission levels. The implementation of mitigation actions will undoubtedly also be local, decentralized – comprised of many individual decisions – and proceed in parallel with adaptation. But for the purposes of modeling impacts it is necessary to simplify this situation.

**Box 5.2** (continued)
adaptation potential and decisions. These include different actors, different timescales and often different spatial scales of decision-making. In particular, mitigation of climate change is only meaningful at the global scale (even if there are potentially large local co-benefits), while adaptation is more likely to be designed at the local scale and yields direct local benefits. These differences make it unrealistic to assume coupled optimization of both policies will occur in any one location let alone worldwide.

Further, there are time horizon differences. Only proactive or anticipatory adaptation, such as changing urban forms, can be designed in conjunction with mitigation policies and strategies, and these represent only a fraction of any aggregate adaptation response. For many adaptation options, decisions about design and implementation will only be made long after initial mitigation choices, when climate actually changes. For these reasons, climate change impact analysis at the local or city scale can consider the global mitigation scenario as an exogenous input.

Finally, assuming "optimal" adaptation does not consider the barriers to adaptation and risk management (see Sect. 3.4) and this can make a large difference in terms of residual impacts. As a result, the best decision on mitigation depends on whether we assume a theoretical world of perfect adaptation, no adaptation, or a world that resembles the current one with some imperfect level of adaptation.

In this methodology (see Fig. 5.3), mitigation benefits at local scale can be assessed across the three distinct adaptation scenarios proposed above, which

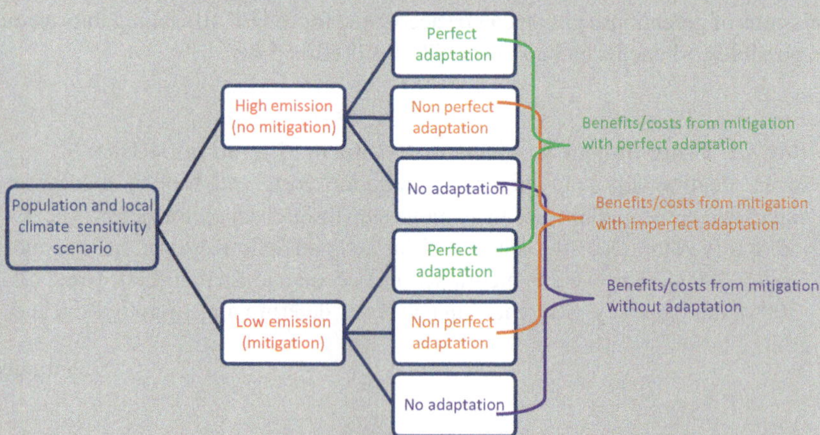


**Fig. 5.3** Assessing the benefits from mitigation

(continued)

**Box 5.2** (continued)
represent a continuum of different possibilities: no adaptation, imperfect adaptation (inspired by observations of the current situation), perfect adaptation. The no-adaptation and perfect-adaptation cases bracket possible outcomes; the scenario inspired from the current observation provides an idea of climate change costs if the world capacity to manage risks does not change significantly in the future.

Unlike global scale analyses, local studies allow us to investigate the interface between global mitigation and local adaptation. Indeed, at the local scale, such as in cities, local mitigation and adaptation use the same policy levers (e.g., urban planning, building standards). Working at local scale also allows researchers to account for important local contextual issues in policy analysis (e.g., the influence of local geographies and development patterns on impacts; or the distribution of institutional authority between local and national actors in key sectors affected by climate change (e.g. see Corfee-Morlot et al. 2011).

This chapter now proposes a few case studies, to illustrate how this methodology can be applied.

## 5.2 Case Study: Hurricanes and the U.S. Coastline[4]

Because of the lack of data on past hurricanes, empirical evaluations of the statistics needed for risk management are very uncertain, even for the past or present climate. To overcome this problem, modeling approaches can be useful.

Here, the Emanuel (2006) hurricane model is used to produce large sets of 3,000 synthetic hurricane tracks, that reproduce the spatial and intensity distribution of hurricanes in the present climate, in the North Atlantic. Because the model can link the spatial distribution of hurricane tracks and intensities to the large-scale climate parameters that are produced by climate models (e.g., mean sea surface temperature, large-scale air circulation patterns), it can be used to assess how climate change may affect hurricane risks. As such, the hurricane model can be understood as a downscaling tool (see Sect. 4.2.2).

A validation of the model in the current climate is proposed in Fig. 5.4, using the HURDAT database, which records hurricane tracks in the North Atlantic from before 1900. We use the 570 tracks that were recorded between 1950 and 2004, because data quality is lower in the first half of the twentieth century. We compare average landfall probability (for location and intensity) calculated with the 3,000

[4]This section is based on Hallegatte (2007a).

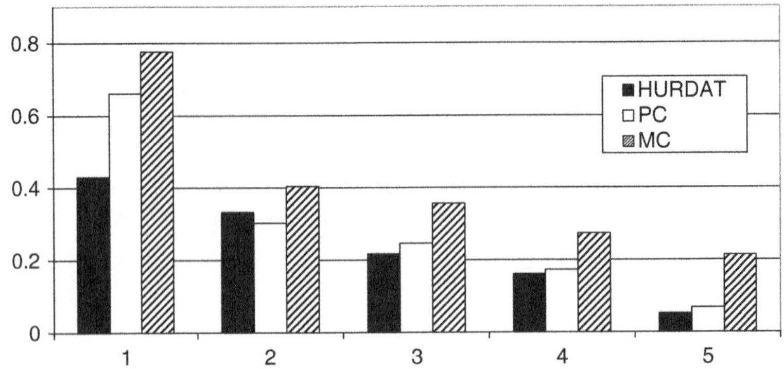


**Fig. 5.4** Annual probability of landfall of a hurricane of a given category, according to historical data (HURDAT), and synthetic tracks in the present (PC) and modified (MC) climate. Climate change effect on hurricane direct losses

synthetic tracks produced by the model and the 570 historical tracks. This validation shows that the model reproduces well the annual probability of landfall of hurricane of categories 2–5, but overestimates the probability of category-1 hurricane landfall. An in-depth analysis shows that this problem is linked to the track of weak hurricanes, which have a bias toward the U.S. coast in the model. This problem is probably due to the fact that weak hurricane storms are not well structured, explaining why the "beta advection" model (the track model in the Emanuel model) is less suitable to assess their tracks and has lower skill than statistical models. This validation suggests that – within some limits – these synthetic tracks can be used to assess risks and climate change impacts.

## *5.2.1 The Hazard: Climate Change and Hurricanes*

This model can then be used to investigate the future of hurricane hazards, corresponding to STEP 2 in the methodology of Fig. 5.1. The model is run with a 10 % increase in potential intensity, which corresponds roughly to a 2°C increase in sea surface temperature (everything else being unchanged), and which is referred to as a "Modified Climate" (MC).

In such an environment, according to the model, annual landfall probabilities increase in all regions, especially for the strongest hurricanes (see Fig. 5.4).

To assess the effect on direct losses (i.e. STEP 4 in the methodology), the vulnerability of each U.S. coastal county is then calibrated using data on past hurricanes and their normalized economic losses. We investigate here the relationship between economic losses and maximum wind speed at landfall. Since population and economic growth are the main long-term drivers of hurricane losses, we use normalized losses, in which the effect of these two drivers has been removed. Such normalized losses have been produced for the 1900–2005 period by Pielke et al. (2008).

We assume that a simple relationship links maximum wind speed at landfall and normalized economic losses:

$$L = \alpha(s)\ W^3,$$

where L measures the normalized economic losses (in millions of U.S.$); W is the maximum wind in $ms^{-1}$; s is the linear coordinate along the coast and $\alpha$ is a parameter that measures the local vulnerability at the location s. This function shape is supported by the shape of the amount of energy dissipated by hurricanes (see Emanuel 2005) and is roughly consistent with empirical damage functions. One shortcoming of this function is that economic losses can rise indefinitely as wind speed increases, while actual damages are limited, in each location, by the total value of assets at risk (see Nordhaus 2006). This relationship is however very conservative compared with other analyses: the statistic analysis by Sachs suggests a much larger value of 6.3; Howard et al. (1972) suggest a mean value of 4.36; and Nordhaus (2006) cites values between 4 and 9. Of course, the larger this value, the larger is the sensitivity of direct losses to a change in hurricane intensity, and our analysis can be considered as conservative on this point.

The parameter $\alpha$ is calibrated for each of the 106 counties along the U.S. eastern and Gulf coast. To do so, we select, for each county, the set of hurricanes that made landfall in this county. When this set is not empty, we fit the parameter $\alpha$ that links maximum wind at landfall and normalized recorded losses (through the minimization of the squared differences). In counties where no hurricanes made landfall (13 counties), the vulnerability is taken as the average vulnerability of all the counties for which we have data. Note that what we call "the vulnerability of county A" can be better defined as "the vulnerability of the U.S. economy to a hurricane making landfall in county A", since this vulnerability also takes into account the losses that occurred in other counties after a landfall in county A.

To validate the ability of this method to assess hurricane losses, these vulnerabilities were applied to the synthetic tracks produced for the present climate (PC) to assess, for each synthetic hurricane that makes landfall, the corresponding economic losses. The mean economic losses are estimated at $1,578 million U.S. per landfall. If we include hurricanes that do not make landfall in the calculation, the mean economic losses per hurricane are estimated at $980 million U.S. per track.

These values are slightly less than historical economic records (respectively $1,833 and $885 million U.S.). The reasons for this discrepancy could be: (i) inaccuracy in the hurricane track and intensity models; (ii) the lack of vulnerability data in some place; (iii) the role of other factors like rainfall and early warnings; (iv) a county is too large as a geographical unit to assess homogenous vulnerability; and (v) historical series are only one realization of the random process and that recorded damages differ from mean damages.

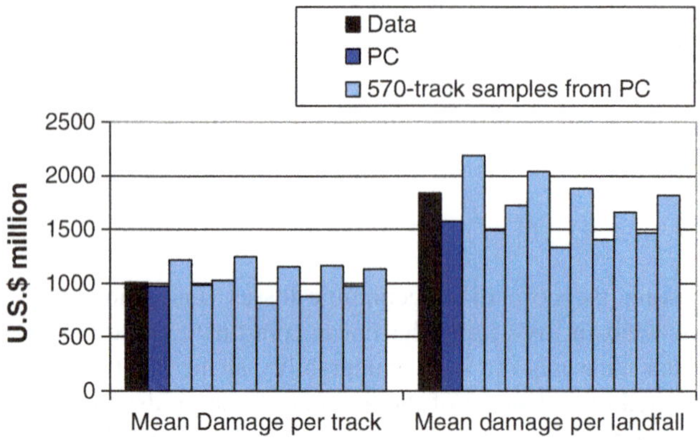


**Fig. 5.5** Mean damage per track and per landfall from historical data, from synthetic tracks generated for the present climate (PC), and from the ten 570-track samples extracted from the 3,000 Present-Climate synthetic tracks

To assess to what extent the difference is a statistical artifact due to the sampling effect, we extract 10 samples of 570 tracks (the same number than historical tracks) from the 3,000 tracks produced for the present climate. These 10 samples can be considered as 10 possible realizations of our model. We then calculate the mean damages for the 10 samples. The mean damages per track and per landfall are reproduced in Fig. 5.5. This figure shows that observed damages are within the range predicted by the model (i.e. with the range of the 10 samples). There is no statistically significant difference between the model and the observations.

When potential intensity is increased by 10 %, the mean economic losses are estimated at $2,375 million U.S. per landfall and $1,514 million U.S. per track, compared with $1,578 and $980 million for the synthetic tracks PC (present climate). Looking back at Fig. 5.5, we observe that such values are outside the range of our 10 samples for the present climate, suggesting that a 10 % increase in potential intensity would produce a change in mean economic losses that would be easily detectable and outside the current natural variability.

**Assuming (i) that the bias in our methodology is unchanged in the future, (ii) that no adaptation is undertaken and (iii) that no change in hurricane genesis takes place, a 10 %-increase in potential intensity is found to cause a 54 % increase in the mean normalized economic losses due to hurricane landfall on the U.S. Atlantic and Gulf coast. Since observed annual damages from hurricane are currently about $8 billion, such an increase would translate into hurricane losses amounting to, on average, $12 billion a year. Of course, this assessment assumes that no socio-economic change takes place in parallel. To assess future hurricane losses, economic and population growths need to be added to this 54 % increase.**

Considering that climate change is expected to make potential intensity increase significantly, this result suggests the possibility of a large increase in hurricane risks in the future. It has to be mentioned, however, that climate change will also influence other environmental parameters, with various effects (see Emanuel 2006): The expected increase in vertical wind shear could reduce mean hurricane intensity; the expected change in ocean thermal structure may, on the other hand, increase hurricane intensity; and a significant sea level rise would make vulnerability rise in low-lying areas (see also the case study in Sect. 5.3). Also, the change in normalized losses will translate into total economic losses that will depend on other factors like population and asset values in coastal areas and on prevention measures that can be implemented (e.g., land-use management, improved building norms, warning and evacuation systems, flood protection systems).

This 54 % increase in the mean annual normalized loss is significant, but it does not seem really threatening for this region, which is one of the richest in the world. This change, nevertheless, should cause an equivalent rise in insurance premiums that can create local issues of insurance affordability (e.g., see RMS 2006, on the insurability of New Orleans).

### *5.2.2 Exposure, Vulnerability and Resilience: Climate Change and Hurricane Losses*

To assess the impact on total losses (i.e. STEP 5 in the methodology), a polynomial regression is estimated on the relationship between direct and indirect losses calculated for Louisiana and Katrina, see Chap. 2. We assume then that the relationship between direct and indirect losses estimated for the state of Louisiana remains valid for all coastal states of the US. This assumption is a strong assumption, and is likely to overestimate indirect losses, because many other states appear more able to cope with large damages than Louisiana. For instance, Florida has a construction sector that is much larger than the Louisiana one, and is able to absorb more easily an increase in demand.

Using this simplifying assumption, we can then assess the total (direct + indirect) losses due to the 3,000 synthetic tracks created by the Emanuel's model. In the present climate, averaged direct losses were estimated at 1,578 million U.S.$ per landfall and 980 million U.S.$ per track. These values translate into averaged total losses estimated at 1,426 million U.S.$ per landfall and 885 million U.S.$ per track. These values are lower than direct losses only, because the economic system is able to limit total losses and make them lower than direct losses for the weakest hurricanes, which constitute the large majority of hurricanes. Indeed, indirect losses are positive only for hurricane causing direct losses in excess of $40b, and only 8 landfalls cause such losses in the 3,000 synthetic tracks created for the present climate. The taking into account of indirect effects leads, therefore, to a minimization of the mean annual economic losses due to hurricanes. In the modified climate, since hurricane intensity is increased according to the Emanuel's model, total losses are larger than in the present climate: the model estimates averaged total losses at 2,272 million U.S.$ per landfall and 1,448 million U.S.$ per track. **When considering indirect losses, the Emanuel's model suggests, therefore, that a 10 % increase in potential intensity would translate into a 59 % increase in total economic losses.**

Possible changes in the frequency of extreme hurricanes, however, are more worrying: among the 3,000 tracks produced by the hurricane model in the present climate, only 59 cause direct losses exceeding $10 billion, and only 4 exceeding $50 billion. Among the 3,000 tracks produced for a climate in which potential intensity has been increased by 10 %, as many as 99 cause direct losses above $10 billion, and 10 cause direct losses above $50 billion. Taking into account indirect losses, a 10 % increase in potential intensity is found to double the likelihood of a hurricane landfall causing more than $50 billion of damages in the U.S. **More than the change in annual losses, therefore, it is the doubling of catastrophic event frequency that could be the most problematic for society and the economy.**

### *5.2.3 Adaptation Options*

From this analysis, an identification of adaptation policies is possible, looking at measures to reduce direct losses (i.e. STEP 7 in the methodology) and indirect losses (i.e. STEP 8). Here, no assessment of their costs and benefits can be proposed. Doing so would require more detailed modeling techniques. Therefore, we only review the options that are available to decision-makers to reduce risks and adapt to climate change.

Numerous actions have been undertaken in the last 100 years to reduce hurricane direct losses and these actions demonstrate that adaptation can be effective. First, there are investments in new protection infrastructures like flood protection systems, dams and building elevations. For example, the sea wall built in Galveston in the aftermath of the 1900 hurricane largely mitigated the consequences of Alicia's landfall in 1983. On a smaller scale, the town of Belhaven, Florida, in the aftermath of hurricane Fran (in 1996), implemented a program to elevate 379 houses. When hurricane Floyd struck in 1999, damages were reduced by 80 %, thanks to this program (Williams 2002).

Second, building codes have been improved and they have limited hurricane damages. The best example is the building code implemented in Florida after Andrew's landfall, even though its effects are only starting to be visible. Also, existing norms have been enforced more rigorously, since hurricanes have shown that the lack of compliance with existing rules had significantly increased damages. The impact of building codes is far from negligible: according to Ryland (2002), "*if all buildings in South Florida were either retrofitted or in compliance with the post-Andrew South Florida Building Code, or the new Florida Building Code that went into effect this year [2002], another Hurricane Andrew would cause only about half* *as much damage to residences and 40 percent less damage to commercial property. The combined loss reduction would be about $10.4 billion, according to the study*".

Third, hurricane track forecasts have improved and better warning systems have been implemented to help people and business to prepare for hurricane landfalls and avoid damages. With early warning, people and businesses can protect houses and suspend dangerous industrial processes, which in turn will reduce direct and indirect damages. For instance, flood damages experts quoted in Carsell et al. (2004) estimate that a 48-h warning can reduce flood damages by up to 50 %, thanks to small-scale preparation efforts inside houses. Also, protecting windows reduces wind damages by 12–54 % (Williams 2002). Preparing industrial facilities can also avoid large pollution and other ancillary damages that can be deadly and costly (e.g., RMS 2005). Most importantly for future vulnerability, much can be done to reduce risk through urban planning and land-use management (Burby and Dalton 1994). Including risk management in land-use planning could very efficiently reduce vulnerability to hurricanes and therefore reduce the direct losses from hurricane, especially where new development is currently occurring.

Indirect losses can also be reduced thanks to policies that increase economic resilience, i.e. improve the ability of the economy to cope with a disaster in an efficient manner. Reconstruction must be done as fast as possible to restore production and housing. For instance, utility companies and the institutions in charge of transport infrastructure must be equipped to face large-scale disasters and reduce as much as possible the period during which their production is interrupted or unreliable. Also, public administrations can facilitate reconstruction, for instance by making it easier and faster to obtain building permits or by allowing workers for other regions to move to the affected area.

## 5.3 Case Study: Sea Level Rise and Storm Surges in Copenhagen

Another case study was carried out in Hallegatte et al. (2011b), on the city of Copenhagen, an industrialized-country low-growth coastal city, and considers the risks from storm surges caused by extra-tropical storms. Because of the high uncertainty on future sea level rise, this analysis assesses flood risks for a large range of possible sea level rises, from 25 to 125 cm.

The case of Copenhagen is interesting because it investigates a city that can be considered as a low-vulnerability high-adaptive-capacity location. Indeed, the city of Copenhagen is today very well protected against storm surge, both by its natural location and by its artificial flood defenses.

The analysis follows the framework presented in Chap. 3: we investigate first the hazards and their probability, then turn to the exposure and sensitivity (or vulnerability), and finally consider resilience and indirect consequences.

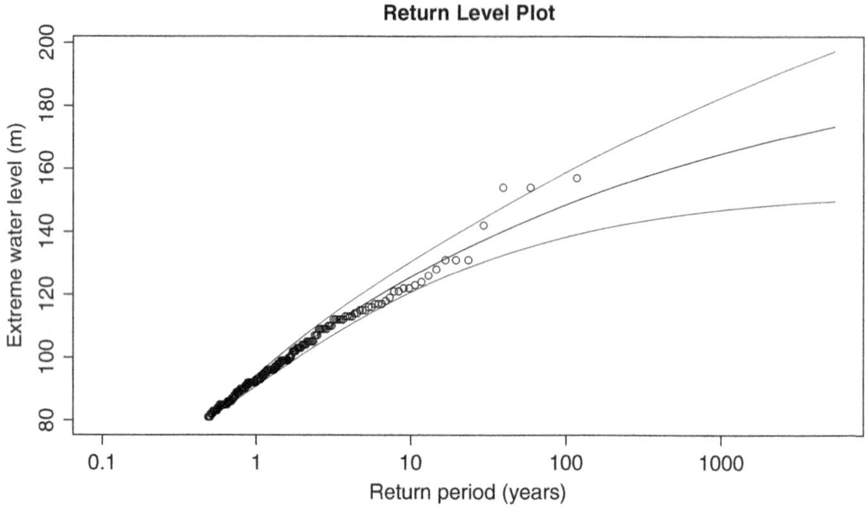


**Fig. 5.6** Storm surge return water level (cm) corresponding to various return-periods, up to 1,000 years. Note: The 117 years of data are reproduced with circles. The presented data was de-trended for extreme analysis

## *5.3.1 The Hazard: Extreme Sea Levels in Copenhagen*

The analysis starts with an analysis of past water levels.[5] Figure 5.6 shows the estimated storm surge water level corresponding to various return periods. Return periods are a measure of frequency; for instance, if the 100-year flood is at 150 cm above normal sea level, it means that there is a 1/100 = 1 % chance of having a water level higher than 150 cm every year. This analysis uses local sea level data (1890 to 2007) provided by the Danish Coastal Authority[6] (Kystdirektoratet) and produced by the Danish Meteorological Institute (DMI). The data set contains 256 events during which water level was at least 80 cm above the mean sea level. Relative frequencies of high-water events of various amplitudes are estimated using a Peak Over Threshold (POT) approach (See Coles 2001). According to this analysis, the water level of the 20-, 50- and 100-year return period events are, respectively, 135 cm, 145 cm and 150 cm.

[5]This analysis has been carried out by Olivier Mestre.

[6]We would like to thank Carlo Sørensen for his help in obtaining this data.

### 5.3.2 The Exposure: Population and Assets at Risk

The second step is to analyze the exposure in the city of Copenhagen, thanks to geographical analysis of population and assets in the city. Population data for Copenhagen was obtained from Landscan 2002 (constrained within metropolitan extent limits) and verified against UN 2005 population data. The metropolitan area was defined using postcode information (from RMS[7] data). Data on residential, commercial and industrial exposure was based on the industry exposure data from the RMS European Winterstorm Model. The exposure of population and asset value to coastal inundation was modeled through geographical mapping onto a Digital Terrain Model (DTM). The DTM was generated from the 90 m (horizontal resolution) elevation database of the Shuttle Radar Topography Mission (SRTM). Exposures by elevation were then extracted from the DTM at 1 m vertical intervals. Finally, cumulative exposures were calculated for each of the sea level events, through interpolating linearly between the vertical intervals (Fig. 5.7).

### 5.3.3 The Vulnerability: Flood Direct Losses

The information on water level statistics and population and asset exposure can then be combined with *vulnerability curves* to assess potential flood direct losses (see Fig. 3.3). Vulnerability curves give the level of damage as a function of flood depth, where the level of damage is defined as the ratio of damages to the total exposure.[8] Since exposure data (as often) does not include infrastructure and public assets, the corresponding losses should be treated separately. Here, it is simply assumed that infrastructure losses correspond to about 40 % of total losses, and they are added as a simple scaling factor.

### 5.3.4 The Resilience: Direct and Indirect Losses

The ARIO regional model (see Chap. 2) is then used to assess the total losses corresponding to different amount of direct losses, i.e. different water levels, i.e.

[7]RMS is Risk Management Solutions, Inc., a private business that provides products, services, and expertise for the quantification and management of catastrophe risk.

[8]Illustrative vulnerability curves have been provided by Risk Management Solutions. It should be noted that early warning, disregarded here, can help reduce content losses. For instance, valuable equipment and furniture can often be saved by households if the event is forecasted early enough. The vulnerability data used here, however, take into account the average effect of early warning and mitigation measures. In the future, improvements in the ability to forecast storms and storm surges may reduce content losses significantly.

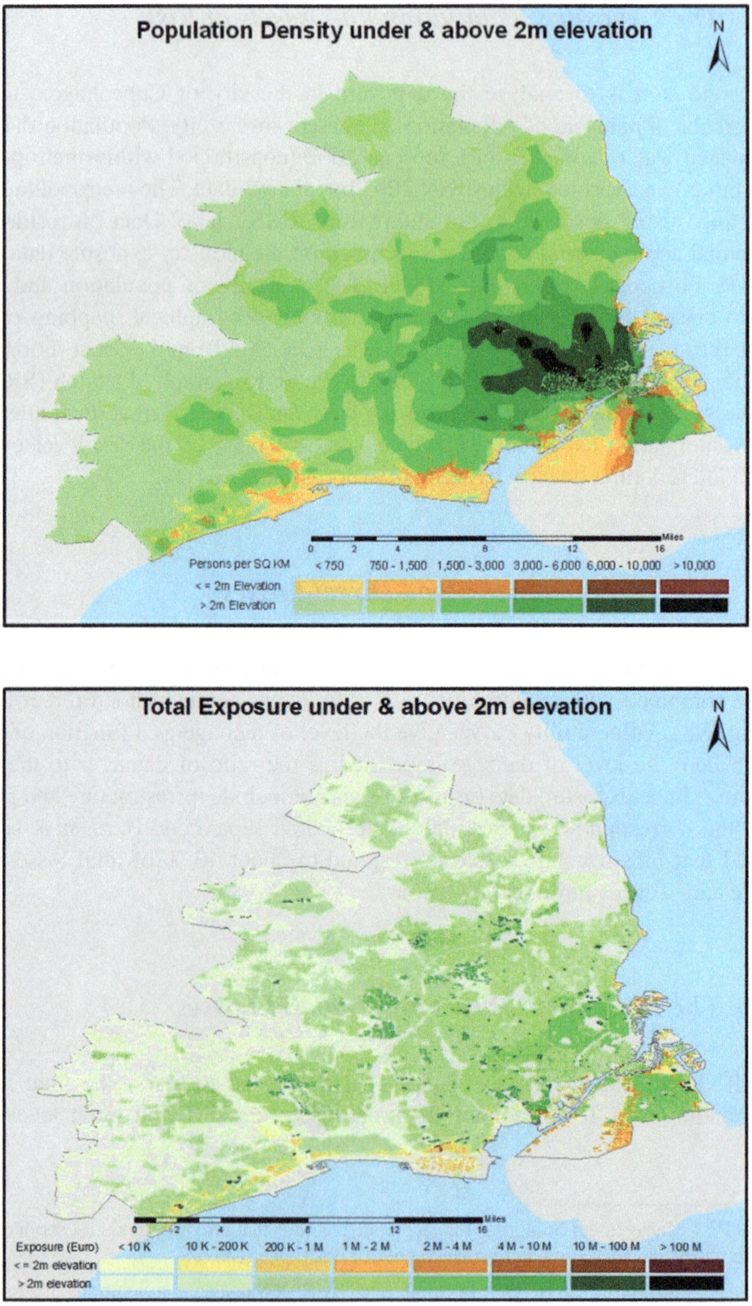


**Fig. 5.7** Population density (*top panel*) and total asset exposure (*bottom panel*) situated in areas with an elevation below (*orange*) and above (*green*) 2 m elevation above sea level

different return periods. Table 5.1 provides an illustration of these losses for various water levels. The analysis reaches the conclusion that, in absence of these defenses, total flood losses would reach today EUR3.1 billion (and about 5,000 jobs) for the 120-year return period event (at 150 cm above average sea level).

From this analysis, sea level rise can then be included, assuming it does not change storm surge statistics. For instance, for no sea level rise, the water level with a 50-year return time would cause total losses amounting to EUR 2.9 billion with no protection. With a 50 cm sea level rise, these losses increase 60 % to EUR 4.7 billion. For the 100-year event, losses are estimated at EUR 3.1 billion in the absence of sea level rise, whereas losses also increase 60 %, to EUR 5.0 billion with a 50 cm sea level rise, and 140 % increase to EUR 7.4 billion with a 1 m SLR (see Fig. 5.8).

### *5.3.5 Adaptation Options*

This type of analysis also allows the comparison of the impacts of sea level rise (in absence of adaptation) and the cost of adaptation (to cancel all impacts) for various amplitudes of sea level rise. This is shown in Fig. 5.9, which shows the mean annual losses in Copenhagen as a function of the level of protection (in cm), for the current sea level and with 50 cm of sea level rise.

Assuming as an illustration that the city is homogenously protected by dikes at 180 cm above current mean sea level, the vertical arrow shows the cost of 50 cm of SLR. This is the increase in mean annual losses caused by a 50 cm SLR in the absence of adaptation (i.e. with no change in the 180 cm protection level).

**This analysis shows that climate change can have threshold effects on natural risks: assuming a 180-cm protection level, a 50 cm sea level rise would make mean annual losses increase from about €300,000 per year to €316 million, i.e. a 1,000-fold increase. This result is due to the fact that current protections have been calibrated using the historical sea level and storm likelihoods; a small change in climate conditions may lead to a very large ill-adaptation of this infrastructure, and to large economic impacts. This result also illustrates the difficulty in estimating climate change impacts, which will depend on small-scale mechanisms like the one described here.**

The horizontal arrow in Fig. 5.9 shows the need for adaptation, i.e. by how much the protection level should be increased to maintain unchanged the mean annual losses due to coastal floods. Using dike cost estimates, this need for adaptation can be translated into adaptation costs.

Maintaining the risk level if sea level rises by 50 cm requires an increase in protections by more than 50 cm. Here, the required increase is equal to 53 cm. Indeed, increasing protections by 50 cm would maintain the probability of a dike overtopping. But in case of overtopping, there are more population and assets affected if dikes are 50 cm higher, and losses are larger. **In other terms and using the framework developed in Chap. 3, higher dikes maintain the hazard unchanged, but lead to a larger exposure. With unchanged vulnerability and**

**Table 5.1** Components of the total flood losses, as a function of water level above current mean level, in absence of protection

| Event sea level wrt current mean level (m) | Direct losses (million EUR) | Indirect losses: Value added losses (million EUR): Loss in productive sectors | Indirect losses: Value added losses (million EUR): Loss in housing services | Indirect losses: Value added losses (million EUR): Total value added losses | Total Losses (direct + indirect) (million EUR) | Lost jobs after 3 months (thousands) |
|---|---|---|---|---|---|---|
| 1 | 1,668 | −58 | 72 | 14 | 1,682 | 3 |
| 2 | 4,837 | −95 | 257 | 162 | 4,999 | 7 |
| 3 | 9,341 | 64 | 682 | 747 | 10,088 | 14 |
| 4 | 14,478 | 517 | 1,446 | 1,964 | 16,442 | 21 |

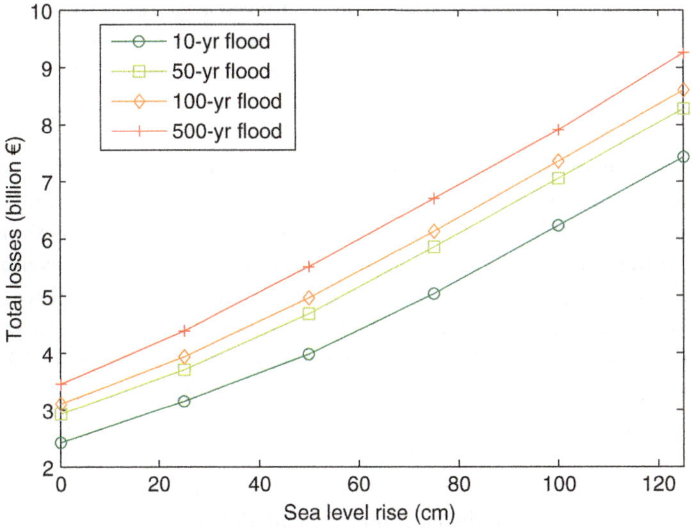


**Fig. 5.8** Total losses caused by the flooding of Copenhagen, as a function of the rise in mean sea level, and for various event return times, in absence of protection

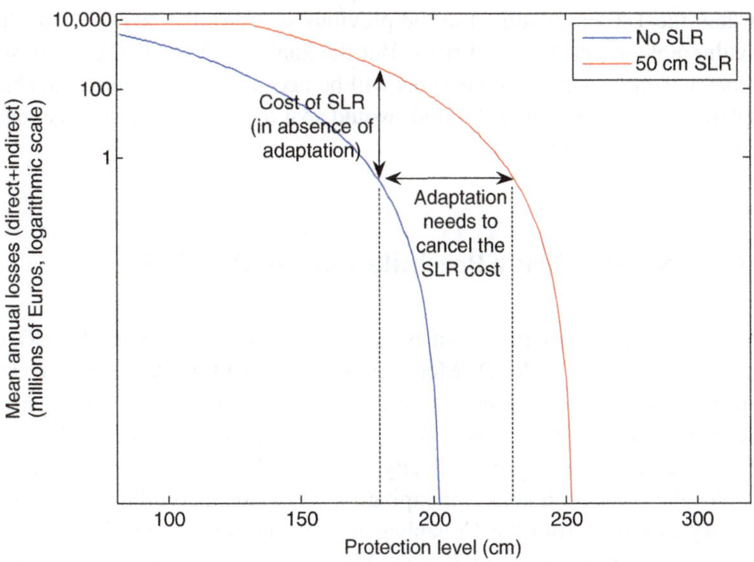


**Fig. 5.9** Illustrative example assuming a homogenous protection at 180 cm above current mean sea level (in the 'No SLR' and '50 cm SLR' cases). The *vertical arrow* shows the cost of SLR in absence of adaptation. The *horizontal arrow* shows the need for adaptation to maintain mean annual losses unchanged

**hazard, an increase in exposure leads to an increase in risk. To maintain the risk level, therefore, the disaster likelihood (i.e. the hazard) needs to be reduced.** The application of the same methodology to 136 cities in Hallegatte et al. (2013) shows that these results – obtained on Copenhagen – are robust to different contexts and environmental and socio-economic conditions.

**This mechanism is met regularly for climate change adaptation and risk management: when climate change forces us to reinforce protections, maintaining the disaster likelihood increases the consequence of a protection failure or overtopping. Climate change makes economies and societies become more vulnerable to exceptional or unexpected events or to protection failure, i.e. less *robust*. This is why Chaps. 6 and 7 focus on methodologies able to take into account the worst-case scenarios and the possibility of errors and failures in the decision-making on risk management, including robust decision-making.**

Figure 5.9, therefore, shows both the cost of SLR in absence of adaptation, and the cost of adaptation to cancel SLR impacts. These cases are two specific options, but other possibilities exist: for instance, one can decide to upgrade protection so that annual mean losses are reduced compared with the current cases (i.e. to do more than adaptation). One can also do less than full adaptation and accept an increase in residual loss. This figure can be used to carry out cost-benefit analysis of coastal protection upgrades in a climate change context.

Thanks to adequate protection levels today (in most of the city, protection is even higher than the 180 cm assumed in the previous section), the city of Copenhagen is not vulnerable to current flood risks. But the analysis shows that even in such a well-protected city, adaptation measures will be necessary in the future. In absence of adaptation, losses from coastal floods would soar to unacceptable levels: inaction is thus not a realistic option.

## 5.4 Case Study: Heavy Precipitations in Mumbai

The last case study was carried out by a consortium including the OECD, Risk Management Solutions, CIRED, Météo-France, NATCOM PMC, and the Indian Institute for Technology Bombay at Mumbai. It was published in Ranger et al. (2011). It investigates the vulnerability to heavy precipitations of the city of Mumbai, a rapidly-growing coastal city in a developing country. The analysis investigates the risks from heavy precipitation and associated urban flooding. This case is very different from the Copenhagen one: the city is considered as highly vulnerable, even in today's conditions. Moreover, the city has lower adaptive capacity, with about 50 % of the population living in slums and a drainage system that is insufficient to deal even with normal monsoon rainfall.

### 5.4.1 *The Hazard: Heavy Precipitations and Extreme Run-offs in Mumbai*

There is no consensus on how precipitation will change over India in response to global warming, with half the models projecting an increase, and half predicting a decrease. The analysis carried out in this project is based on one climate model, the Precis model forced by the HadCM3 model in a SRES A2 scenario. Since this model projects a large increase in precipitation over India, the results in terms of flood risks can be considered as falling within the upper range of future risks.

This analysis starts from an analysis of past high precipitation events, and assesses the impact of climate change on heavy precipitation statistics using downscaling techniques (namely, the WXGEN weather generator). Then, it translates the change in the statistics of heavy precipitation into river run-off and flood hazards, using the Storm Water Management Model (SWMM).

This exercise allows producing flood maps, in the current climate and in the climate projected by the PRECIS climate model with the IPCC/A2 emission scenario (Fig. 5.10). These flood maps are highly uncertain and depend on the selected climate model. But they provide a first piece of information useful to help urban planners identify where additional development may reveal risks and dangers in the future.

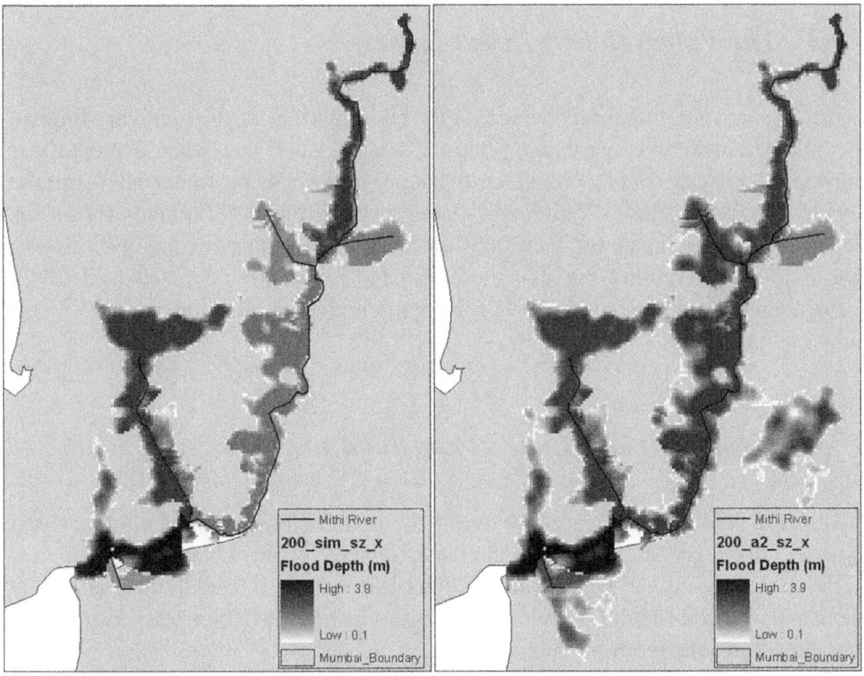


**Fig. 5.10** Flood map corresponding to the 200-year return period precipitation event, in the Mithi basin, in Mumbai, today (*left panel*) and in the 2080s in one climate scenario (*right panel*)

### 5.4.2 The Exposure: Population and Assets in Mumbai

Then, we can assess the population and assets exposed to increased risks, using data on population and assets collected by Risk Management Solutions from an insurance database developed for the assessment of earthquake risks, and assuming that the city remains unchanged until the 2080s.

Of course, this assumption of a static city is for illustrative purpose only, to provide an estimate of how much climate change can make flood risks increase in the future. A population map (on a 100 m grid) was developed from publicly available 2001 census data (at the ward level). The distribution of residential, commercial and industrial property types was derived from observations by the IRS LISS III satellite (Indian Remote Sensing Satellite, Linear Image Self Scanning III) and panchromatic images at a resolution of 10 m. Six exposure types were defined: two residential (low density and high density), three commercial (low-rise retail and offices, high-rise office blocks and skyscrapers) and one industrial. The total insured values (TIVs) of insured assets are based on the RMS India Earthquake Model (INEQ). The TIV can be converted to a total value if the insurance penetration is known. Here, the insurance penetration is estimated at 8 % for residential properties, 14 % for commercial properties and 17 % for industrial properties. There is a large uncertainty on these estimates.

### 5.4.3 The Vulnerability: Direct Losses

In the absence of vulnerability curves for the buildings that can be found in Mumbai, the analysis uses "average damage ratio". It is assumed that when a property is flooded, a constant share of its value is lost, regardless of the water level and the detailed characteristics of buildings. Using three different techniques (based on published loss estimates for the 2005 floods, insurance data for the 2005 floods, and simple vulnerability curves), these damage ratios are estimated at 5–15 % for residential properties, 15–35 % for commercial properties, and 10–30 % for industrial properties.

### 5.4.4 The Resilience: Indirect and Total Losses

Finally, the ARIO model is used to assess the indirect losses that would be caused by such floods, accounting for the characteristics of the Mumbai region economy. In particular, the model was modified to account for the financial constraints that can make it very difficult for poor households to finance the reconstruction, and can thus slow down the recovery significantly.

**Table 5.2** Upper estimation of total losses (direct + indirect, including loss in housing services) due to various types of events in present-day and future conditions

| | Projected flood losses ($ million USD) | | | | | |
|---|---|---|---|---|---|---|
| | Present-day | | | 2080s | | |
| Type of event | Direct losses | Indirect losses | Total losses | Direct losses | Indirect losses | Total losses |
| Simulated July 2005 | 1,910 | 425 (18 %) | 2,335 | | | |
| 50-year RP | 570 | 95 (14 %) | 665 | 760 | 130 (15 %) | 890 |
| 100-year RP | 600 | 100 (14 %) | 700 | 1,890 | 415 (18 %) | 2,305 |
| 200-year RP | 600 | 100 (14 %) | 700 | 1,990 | 445 (18 %) | 2,435 |

Note: In parenthesis is the contribution of indirect economic losses to the total losses

**Using this model, flood risks are found to increase very rapidly in response to climate change; for instance the total losses (direct plus indirect) caused by the 100-year flood would more than triple, with indirect losses representing between 15 and 20 % of direct losses (see Table** 5.2**).**

### *5.4.5 Adaptation Options*

The paper also investigates the economic benefits of a set of adaptation options (see Fig. 5.11). These include an upgrade in the drainage system and an increase in building quality. **Results suggest that the losses due to a 100-year event can be reduced below their current level in spite of climate change and the corresponding increase in extreme rainfall intensity. This comforting result shows again that an increase in hazard does not need to translate into more disasters, if adequate measures are implemented. In places like Mumbai, where the adaptation deficit is extremely large, progress in risk management could more than compensate climate change impacts (see also Sect. 6.2 on the concept of "adaptation deficit").**

The analysis also considers the role of insurance in reducing welfare losses and increasing reconstruction capacity, therefore reducing indirect losses (STEP 5 of the methodology). For instance, the modeling exercise suggests that increasing insurance penetration from about 10 % today to 100 % would reduce indirect losses by 50 %, because it does not make it necessary for households to reduce their consumption patterns to pay for reconstruction.

This is already a significant impact, but insurance – and micro-insurance in particular – could have an even larger impact on welfare thanks to its role in spreading risks and mitigating distributional effects, especially for the poorest.

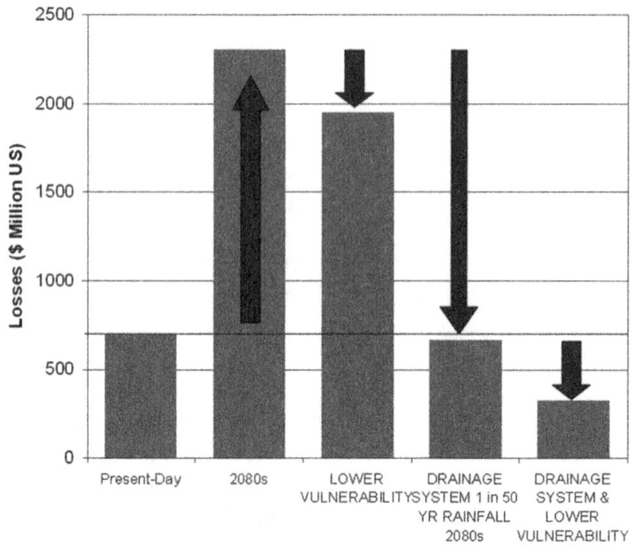


**Fig. 5.11** The estimated total (direct + indirect) losses for a 1-in-100 year flood event in Mumbai under five scenarios (from *left* to *right*): (i) present-day; (ii) 2080s – using the one 'high-end' scenario considered in this study and an unchanged city; (iii) 2080s, assuming properties are made more resilient and resistant to flooding (e.g. through building codes); (iv) 2080s, assuming the drainage system is improved such that it can cope with a 1-in-50 year rainfall event; and (v) combined property and drainage improvements

### *5.4.6 Impact on Marginalized Populations*

Aggregated losses can hide large impacts on some of the population. This is especially important in places like Mumbai, where wealth and living standard differences are very large within the city. As an illustration, this section is an attempt to describe the impact of the 2005 floods on the marginalized population living in disaster prone regions of the city. Data was collected from households through a comprehensive primary survey undertaken in the worst affected regions of greater Mumbai. The survey was conducted by the Indian Institute of Technology Bombay in Mumbai, under the direction of A. Patwardhan and K. Narayanan. More complete results are provided in Hallegatte et al. (2010).

The household size in the survey sample varies from a maximum of 35 members to a minimum of 1 member. The mean of the sample stands at just above six members. The average number of children in the households stands at around two members. About 82 % of the households derive most of their income from the service sector. The primary informal sector activities relates to informal establishments involved in processing of leather products, small lock and key making units, units engaged in making garments and toys and other units manufacturing goods and provisions for day to day living of households in these regions.

The flood witnessed in Mumbai during 2005 was rare in terms of the severity of the event. Although small floods are common in the regions during almost every monsoon, the scale of the 2005 event was not comparable with others. Most of the households say that the event during 2005 was an isolated one and they have never faced such a situation before. Almost the entire sample was affected due to the floods in 2005. Similarly a majority of the households (around 90 %) also believe that the impact of the floods on their family was very high. Only a select few households amounting to only 10 % of the sample reported that the impact of the event on their family was medium. Around 39 % of the households said that they had to reduce their food intake due to the disaster event.

Also, a majority of the households report that they did not receive any financial support from financial agencies after the event nor were able to pay back loan installments due to the event. It is also interesting to note that around 96 % of the households also say that there has been no change in their livelihood due to the floods in 2005. It seems thus that the floods did not have a permanent effect on population livelihood.

Households reported that, during the 2005 flood, they suffered damage to their housing structures and assets. The reconstruction cost on an average works out to be around Rs. 4,700 ($108), and their other asset losses were about Rs. 7,700 ($177). These values are large compared with household resources: the average annual income reported by the households stand at around Rs. 8,400 ($193) and savings stand at around Rs. 6,900 ($159). The total value of their asset is about Rs. 35,000 ($805). So, around 20 % of the household assets were lost due to the flood event during 2005.

Around 97 % of the households depended on their past savings to fund the reconstruction of their damaged houses. Only 0.2 % of the households reported that they took money from money lenders to finance their reconstruction cost. Similarly around 1.7 % of the households say that they have secured the money for reconstruction of their houses from family members and another 0.7 % report of having the money from other sources.

The survey also assessed the total losses to marginalized population at about $250 million in the entire city. Compared with total losses, see Table 5.2, this sum represents only a small fraction of losses. However, the survey shows that in terms of welfare impacts, this fraction of total losses may play a disproportional role. Care should thus be used when assessing disaster risks with aggregated measures only.

**In the case of Mumbai, the impact of climate change on slums and marginalized population may be more of a problem than the total impact on the entire city, even though the corresponding losses are small compared with total losses.** This analysis shows that classical indicators of losses and climate change impacts may hide significant impacts on vulnerable population, and that distributional effects need to be considered.

**These results also suggest that natural disasters and climate change can be a strong obstacle to poverty alleviation.** It is well documented, see for instance the 2014 World Development Report by the World Bank (2013), that building assets is critical for households to escape poverty. Because disasters can swipe assets – as shown by the 2005 flood in Mumbai – they can bring back households into poverty. The fact that the losses to poor households were of the same order of magnitude as their total savings suggests that events such as the 2005 floods are able to bring people back into poverty. And the flood may not even need to occur: knowing that such a flood is possible may lead households and firms not to invest in better housing or more efficient productive equipment. As a result, the risk for a flood – and not the flood itself – may be an obstacle to poverty reduction.

## 5.5 Lessons from the Case Studies

The case studies suggest that the risks from natural hazards may increase rapidly in the future, because of a confluence of factors most notably socio-economic growth and climate change. **Where protections are in place, even a limited climate change can make them inadequate and have a very large multiplying effect on average disaster risk losses.**

However, specific adaptation measures can be implemented to limit flood risks, for instance with sea walls and other coastal protections, improved drainage infrastructure (possibly with a change from gravity drainage to pumps), generalized insurance systems, early warning and evacuation systems, and enhanced reconstruction cases. These measures are meant to reduce either or both direct losses (i.e., destruction from natural hazards) and indirect losses (i.e., the welfare impacts of these hazards). It is likely that an optimal adaptation policy mix includes both measures aiming to reduce direct losses and measures targeting indirect losses (see next chapter on how to design it).

**The implementation of adaptation and risk mitigation measures can maintain future disaster losses at their current level, or even below their current level, especially in high-vulnerability areas. The impact of climate change will largely depend on whether adequate (possibly costly) policies will be implemented to cope with it.**

Although a full economic assessment of adaptation options was not carried out, in many cases measures may be "no regret," that is they would yield benefits if implemented today, even in absence of climate change. This is especially true in developing-country cities, where the adaptation deficit is very large.

Yet they often require significant anticipation to be effective, either because they require a lot of planning and design (e.g., sea walls or flood barriers) or because they are efficient only over the long term (e.g., land-use planning). For these reasons, it would be particularly beneficial to start immediately to develop careful assessments of such adaptation options, with local, context-specific information so as to design and implement adaptation and risk management plans in coastal cities. Such plans should integrate climate change considerations within routine urban land-use and infrastructure planning. This is especially true in developing country cities, because they are growing rapidly and they are in the process of designing and building urban infrastructure, which will determine urban form, and thus to some extent long-term vulnerabilities.

## 5.6 Conclusion on the Future of Natural Disasters and the Role of Climate Change

The current adaptation to natural hazard is far from optimal and there are large potentials for improvement in how we manage disaster risks (2014 World Development Report, World Bank 2013, chapter 2). **Through better risk analyses, and the implementation of better risk management policies, human and economic losses from natural disasters could be largely reduced, sometimes even below their current level in spite of climate change.**

However, there are reasons why such better measures are not implemented, and they are not only financial reasons. Institutional weaknesses and political constraints may prevent disaster risks from being managed in a rational way, increasing disaster losses. Moreover, where climate change increases natural hazards, for instance because of sea level rise or intensified storms or floods, managing risks will be even more complicated, and possibly more expensive. It would be over-optimistic, therefore, to start from the assumption that future risk management will be better tomorrow than what is observed today and that the right response to climate change will be implemented. **It is thus necessary to assess the consequence of climate change taking into account the likely sub-optimality of adaptation responses.**

**The cost of climate change through modified extremes events will be felt by a combination of modified protection costs (the cost of adaptation) and of modified residual damages from disasters (the residual cost of climate change when adaptation measures have been implemented).**

Careful attention should be devoted to the fact that these two costs can increase nonlinearly with climate change:

- Protection costs increase nonlinearly with natural hazards. Where natural coastal protections exist, for instance, sea level rise may make it necessary to replace them with artificial protections. In that case, even a limited sea level rise may result in soaring protection costs. Also, climate change and sea level rise may make it necessary to move from gravity drainage to pump stations in coastal cities, increasing the corresponding cost in a significant manner.
- Residual damages from disasters can also increase nonlinearly with hazards. As illustrated on the Copenhagen case, damages increase rapidly if a well-calibrated protection system becomes ill-adapted to new climate conditions. Also, as seen in Chap. 2, indirect losses amplify this nonlinearity: beyond a coping capacity threshold, indirect losses increase more rapidly than direct losses, and when the reconstruction capacity is exceeded, disasters may lead to a bifurcation toward lower development levels, large GDP losses, or poverty traps.

**Where protection costs and/or residual damages exceed an acceptable threshold, one can imagine that retreat from the affected zones may be the best answer. But such a retreat is impossible in many places, including the large cities where more and more population is living. In these locations where retreat appears impossible, climate change costs may be large.**

**The case of the very poor is also critical**: they are more exposed to risks than the others, disasters represent a larger burden for them (see also Chap. 3, Sect. 3.2 for a global analysis), and risks and disaster losses can be a large obstacle for them to escape poverty. Even if the absolute numbers at stake are not large – because they are so poor – they suffer from most of the welfare impacts. **Aggregated estimates of disaster losses are very bad indictors of the effect of risks and disasters on the poorest, and it is useful to complement them with specific analyses of the impact on poverty.**

## References

Burby RJ, Dalton LC (1994) Plans can matter! The role of land use plans and state planning mandates in limiting the development of hazardous areas. Public Adm Rev 54(3):229–238

Carsell KM, Pingel ND, Ford DT (2004) Quantifying the benefit of a flood warning system. Nat Hazards Rev 5:131–140

Coles S (2001) An introduction to statistical modelling of extreme values. Springer, London, 208 pp

Corfee-Morlot J, Cochran I, Hallegatte S, Teasdale PJ (2011) Multilevel risk governance and urban adaptation policy. Clim Chang 104(1):169–197

Emanuel K (2005) Increasing destructiveness of tropical cyclones over the past 30 years. Nature 436:686–688

Emanuel K (2006) Climate and tropical cyclone activity: a new model downscaling approach. J Clim 19:4797–4802

Fankhauser S, Smith JB, Tol RSJ (1999) Weathering climate change: some simple rules to guide adaptation decisions. Ecol Econ 30(1):67–78

Hallegatte S (2007) The use of synthetic hurricane tracks in risk analysis and climate change damage assessment. J Appl Meteorol Climatol 46(11):1956–1966

Hallegatte S (2009) Strategies to adapt to an uncertain climate change. Glob Environ Chang 19:240–247

Hallegatte S, Hourcade J-C, Ambrosi P (2007) Using climate analogues for assessing climate change economic impacts in urban areas. Clim Chang 82(1–2):47–60

Hallegatte S, Ranger N, Bhattacharya S, Bachu M, Priya S, Dhore K, Rafique F, Mathur P, Naville N, Henriet F, Patwardhan A, Narayanan K, Ghosh S, Karmakar S, Patnaik U, Abhayankar A, Pohit S, Corfee-Morlot J, Herweijer C (2010) Flood risks, climate change impacts and adaptation benefits in Mumbai: an initial assessment of socio-economic consequences of present and climate change induced flood risks and of possible adaptation options. OECD environment working papers, no. 27, OECD Publishing, Paris, France

Hallegatte S, Henriet F, Corfee-Morlot J (2011a) The economics of climate change impacts and policy benefits at city scale: a conceptual framework. Clim Chang 104(1):51–87

Hallegatte S, Ranger N, Mestre O, Dumas P, Corfee-Morlot J, Herweijer C, Muir Wood R (2011b) Assessing climate change impacts, sea level rise and storm surge risk in port cities: a case study on Copenhagen. Clim Chang 104(1):113–137

Hallegatte S, Green C, Nicholls R, Corfee-Morlot J (2013) Flood losses in major coastal cities. Nat Clim Chang 3(9):802–806

Howard RA, Matheson JE, North DW (1972) The decision to seed hurricanes. Science 176:1191–1202

Jamet S, Corfee-Morlot J (2009) Assessing the impacts of climate change: a literature review, No. 691. OECD Publishing, Paris

Klein RJT (2001) Adaptation to climate change in German official development assistance – an inventory of activities and opportunities, with a special focus on Africa. GTZ climate protection programme, 42 p, Eschborn

Lecocq F, Shalizi Z (2007) Balancing expenditures on mitigation of and adaptation to climate change: an exploration of issues relevant to developing countries, 1 Aug 2007. World Bank Policy Research working paper 4299, Washington, DC, USA

Manning M, Petit M, Easterling D, Murphy J, Patwardhan A, Rogner H, Swart R, Yohe G (2004) IPCC workshop report: describing scientific uncertainties in climate change to support analysis of risk and of options. IPCC, Geneva

Nicholls R, Leatherman S (1996) Adapting to sea-level rise: relative sea-level trends to 2100 for the United States. Coast Manag 24(4):301–324

Nordhaus WD (2006) The economics of hurricanes in the United States, prepared for the Snowmass workshop on abrupt and catastrophic climate change. Annual Meetings of the American Economic Association, Snowmass

Pielke RA Jr, Gratz J, Landsea CW, Collins D, Saunders MA, Musulin R (2008) Normalized hurricane damages in the United States: 1900–2005. Nat Hazards Rev 9(1):29–42

Ranger N, Hallegatte S, Bhattacharya S, Bachu M, Priya S, Dhore K, Rafique F, Mathur P, Naville N, Henriet F, Herweijer C, Pohit S, Corfee-Morlot J (2011) A preliminary assessment of the potential impact of climate change on flood risk in Mumbai. Clim Chang 104(1):139–167

RMS (2005) Hurricane Katrina: profile of a super cat. Lessons and implications for catastrophe risk management. https://www.rms.com

RMS (2006) Flood risk in New Orleans: implications for future management and insurability. http://www.rms.com/NewsPress/PR_120406_NOFloodRisk.asp

Rozenberg J, Hallegatte S, Vogt-Schilb A, Sassi O, Guivarch C, Waisman H, Hourcade JC (2010) Climate policies as a hedge against the uncertainty on future oil supply. Clim Change 101:663–668

Ryland HG (2002) Better building codes cut property loss. National Underwriter, 106

Solecki WD, Oliveri C (2004) Downscaling climate change scenarios in an urban land use change model. J Environ Manag 72:105–115

Tol RSJ (2005) Adaptation and mitigation: trade-offs in substance and methods. Environ Sci Policy 8(6):572–578

Tol RSJ, Fankhauser S, Smith JB (1998) The scope for adaptation to climate change: what can we learn from the impact literature? Glob Environ Chang 8(2):109–123

Viguié V, Hallegatte S (2012) Synergies and trade-off in urban climate policies. Nat Clim Chang 2:334–337

Viguié V, Hallegatte S, Rozenberg J (2014) Downscaling long term socio-economic scenarios at city scale: a case study on Paris. Technol Forecast Soc Change 87:305–324

Weitzman ML (2009) On modeling and interpreting the economics of catastrophic climate change. Rev Econ Stat 91(1):1–19

Williams BA (2002) Fran, Floyd and mitigation policy. Berry A. Williams and Associates, Inc, Collinsville, IL, USA

World Bank (2012) Inclusive green growth. The World Bank, Washington, DC

World Bank (2013) World development report 2014. Risk and opportunity: managing risk for development. World Bank, Washington, DC

Yohe G, Neumann J, Ameden H (1995) Assessing the economic cost of greenhouse induced sea level rise: methods and applications in support of a national survey. J Environ Econ Manag 29:S-78–S-97

Yohe G, Neumann J, Marshall P, Ameden H (1996) The economic cost of greenhouse induced sea level rise in the United States. Clim Chang 32:387–410

Yohe G, Knee K, Kirshen P (2011) On the economics of coastal adaptation solutions in an uncertain world. Clim Chang 106(1):71–90

# Chapter 6
# Methodologies for Disaster Risk Management in a Changing Environment

**Abstract** This chapter compares reactive risk management practices and the proactive approaches that are required in a climate change context. Since there is no silver bullet policy against risk, risk management and climate change adaptation needs to rely on a policy mix, from physical protection to land-use planning and insurance. Integrated and proactive approaches are more efficient, but also more difficult to design and implement. The chapter presents a methodology to perform cost-benefit analyses for risk management policy assessment, with two case studies on New Orleans coastal protections and investments in early warning systems in developing countries. It stresses the limits of this approach, especially when uncertainty is very large or when distributional and non-market effects are important.

**Keywords** Disaster risk management • Cost-benefit analysis • Uncertainty • Decision-making

A conclusion from previous chapters is that climate change is very likely to make natural hazards increase in many regions, and that losses could soar in absence of adequate policy response. It showed also that adaptation actions are available to limit or prevent an increase in disaster losses. But implementing the necessary actions will require specific risk management practices and methodologies. This chapter describes methods to assess disaster risk management options, and discusses their ability to cope with climate change.

## 6.1 The Disaster Risk Management "Policy Mix"

There is no single measure that can provide full protection against all natural hazards. The highest and strongest dikes have a probability of failure, and risk can never be fully avoided, by any measure and at any cost. Moreover, some approaches are more efficient in coping with frequent, low-intensity events while others are best at managing rare, high-intensity events. For instance, it is easy to avoid the frequent floods that occur in Mumbai almost every year through improvements in the drainage system. But it is almost impossible to prevent floods in case of



S. Hallegatte, *Natural Disasters and Climate Change*,
DOI 10.1007/978-3-319-08933-1_6

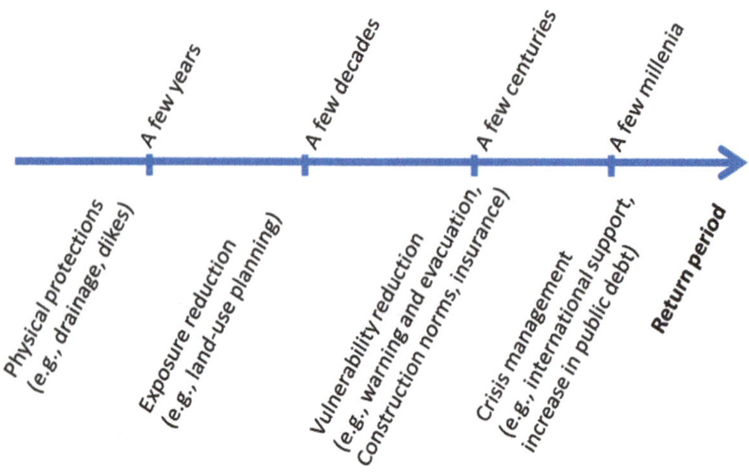


**Fig. 6.1** An example of risk-management policy mix, in which physical protections avoid frequent events, land-use planning limit losses in these protections are overtopped, and early warning, evacuation, insurance and crisis management cope with the largest events

exceptional rainfall like in July 2005 (see Sect. 5.4). To cope with such exceptional events, early warning and evacuation systems can play a very efficient role. But early warning cannot do anything against frequent events: large parts of Mumbai cannot be evacuated several times a year.

**Risk management should be done through a policy mix, with several policies targeting different return periods, as suggested in Fig. 6.1. Using the wording suggested in Chap. 2, some of these policies will target direct disaster impacts and try to minimize them (disaster risk reduction actions), while others will target indirect disaster impacts and try to increase resilience (resilience building actions).**

**To minimize the cost of risk management, and maximize its benefits, the different policies of the mix should be designed together. With very strong physical protections – like in the Netherlands – there are different needs in terms of evacuation system, insurance regulations and building norms. Where financial constraints make such protection unaffordable – like in Bangladesh – it is all the more critical to implement an efficient early warning and evacuation scheme.**

There is no "optimal" risk management policy mix, and different approaches are possible. Rich countries may decide to focus on physical protections, while poor countries may prefer to invest in early warning and evacuation. Depending on which approach is selected, the at-risk areas may be affected more or less often, with more or less damages.

Importantly, a change in one component of the risk management policy mix may require changes in the other components. For instance, if the protection level provided by physical infrastructure is reduced by climate change or subsidence, the

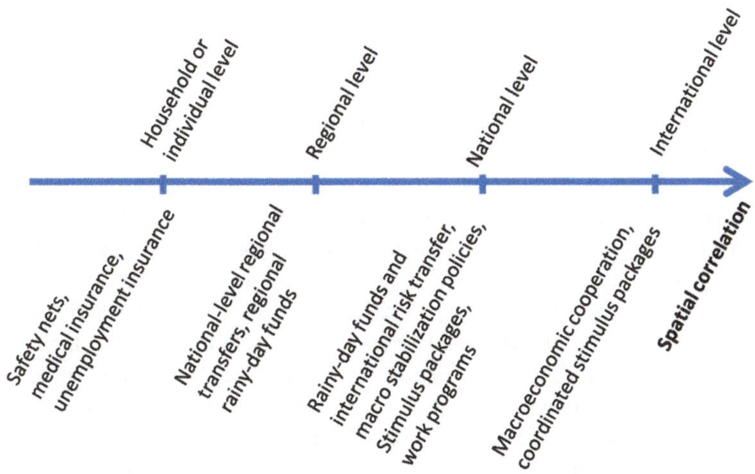


**Fig. 6.2** Policies to cope with correlated risks, depending on the spatial correlation

criteria used to decide when to evacuate at-risk areas may have to be revised; and insurance scheme financial viability may be threatened.

Finally, risk management needs also to be designed as a series of scales (see Fig. 6.2). For idiosyncratic shocks – car accidents or illness – risk sharing across a small population may be sufficient. But when correlation increases – say for instance for floods or epidemics – then a large share of the local population may be affected at the same time, stretching the coping capacity of local systems (Box 6.1). This is when risks need to be transferred to a higher scale, at the regional or national level.

And when risks cannot be shared horizontally – across a larger population in one given year – they need to be shared across time, through increased savings (with non-correlated returns) and reinsurance schemes.[1]

In practice, however, such a risk management approach that takes a holistic and integrated view and coordinates across policy options (Fig. 6.1) and across scale (Fig. 6.2) is made difficult to implement by institutional fragmentation and coordination issues. Developing such a risk-management package requires a concerted effort by the ministry in charge of water and coastal management (for the dikes and drainage system), the ministry in charge of land-use and urbanization and various local authorities (for land-use planning), the ministry of interior (often in charge of warning, evacuation, and crisis management), the ministry of finance (for insurance and emergency finance), and the ministry of foreign affairs (if external support in case of disasters is concerned). It is extremely difficult to organize such a dialogue in a constructive way, and second-best – more fragmented – approaches are often adopted.

[1]Reinsurance firms receive insurance premiums every year and save (and invest) them to be able to pay claims when a disaster hits. As such, reinsurance premiums are a form of savings.

**Box 6.1: The Limits of Individual and Community Risk Management**
The 2014 World Development Report reviews in its Chaps. 3 and 4 the literature on risk management and coping strategies of poor households and communities, and concludes that poor households are particularly exposed to natural risks. To cite an example, a common practice in poor communities is to save in-kind, for instance by advancing the construction of a house whenever resources become available. Such physical in-kind savings are typically subject to the direct risk of destruction in the event of a disaster. Moreover, households attempting to sell off assets (such as livestock) in the aftermath of a disaster often find that other households are also trying to sell their assets, leading to a collapse in local demand and thus of prices (Dercon 2007; Fafchamps et al. 1998; Kazianga and Udry 2006). For the same reason, community-based risk sharing mechanisms (e.g., family networks or informal solidarity mechanisms) often fail in the context of natural disasters: for instance, a local insurance provider is likely to be unable to provide full insurance coverage if all its members and clients are simultaneously affected by a catastrophic disaster (Skoufias and Quisumbing 2003).

To build more integrated risk management strategies, decision-making processes for risk assessment and implementation need to include multiple stakeholders. Doing so is useful not only to disseminate information and increase the political acceptability of selected solutions, but also to enhance the technical quality of the analysis and of resulting strategies. In many countries, multi-ministry bodies are now in charge of information exchange and coordination for disaster risk reduction. The responsibility for risk management is located in the highest government office in about 25 % of the cases. In Peru, for instance, a new agency located in the president's office has the responsibility for disaster risk management.

## 6.2 Disaster Risk Management for Climate Change Adaptation

**Disaster risk management policies can be important tools for climate change adaptation. But to do so, they have to be designed to accommodate for evolving hazards and uncertain trends, and avoid using past events and observations as unique input information.**

### *6.2.1 Reactive vs Proactive Risk Management*

The prologue of this book describes how two regions – the New Orleans region and the Netherlands – have managed changing hazards, with local soil subsidence that continuously increases flood risks. What most locations in the world will experience tomorrow because of global sea level rise and changing climate conditions, indeed, looks like what these two regions have experienced in the last 100 years. Investigating how these regions managed this increase in hazard provides interesting insights into relevant adaptation strategies.[2]

What is striking in the comparison of New Orleans and the Netherlands is the difference in the nature of their response to disasters. In New Orleans, the city responded to floods by implementing ambitious protection upgrades, without any institutional or legal change. In the Netherlands, the response to the 1953 flood was not only an engineering response: a commission, the Delta committee, was created to manage the response from an institutional and technical point of view. Using cost-benefit analyses, the Delta committee determined an acceptable level of flood risk and, from it, derived an optimum level of protection, formulated as the return period of the water level for which protections are designed (see Fig. 6.3). The Dutch Law on Water Defences also requires that water levels and wave heights used in risk analyses and in the design of protections, should be updated every 5 years and that water defenses should be evaluated for these new conditions.

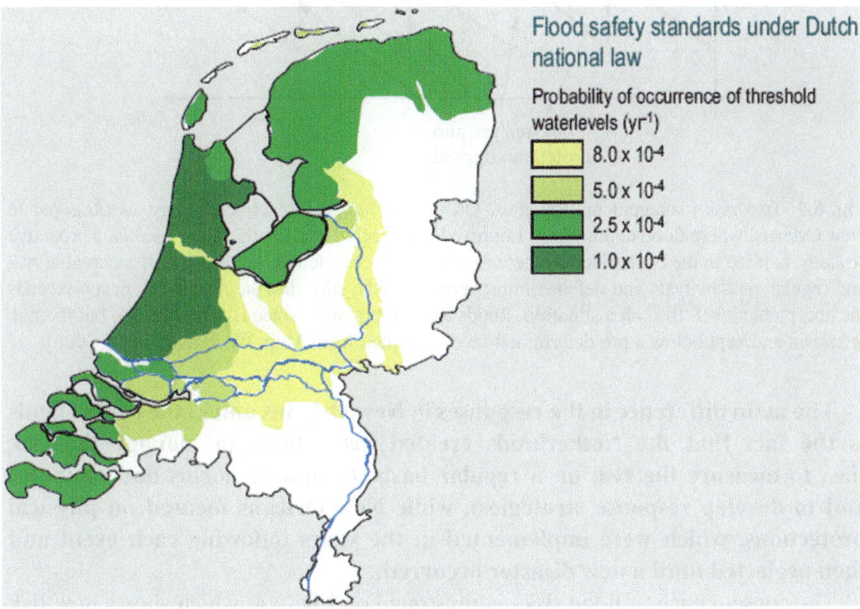


**Fig. 6.3** Flood safety standards under Dutch national law (From Netherlands Environmental Assessment Agency, National Institute for Public Health and the Environment)

[2]Burby et al. (2006) provides a comparison of risk management practices in Florida and Louisiana. Their analysis shows that, even within the U.S., risk management practices differ widely, and lead to more or less losses over the long term.

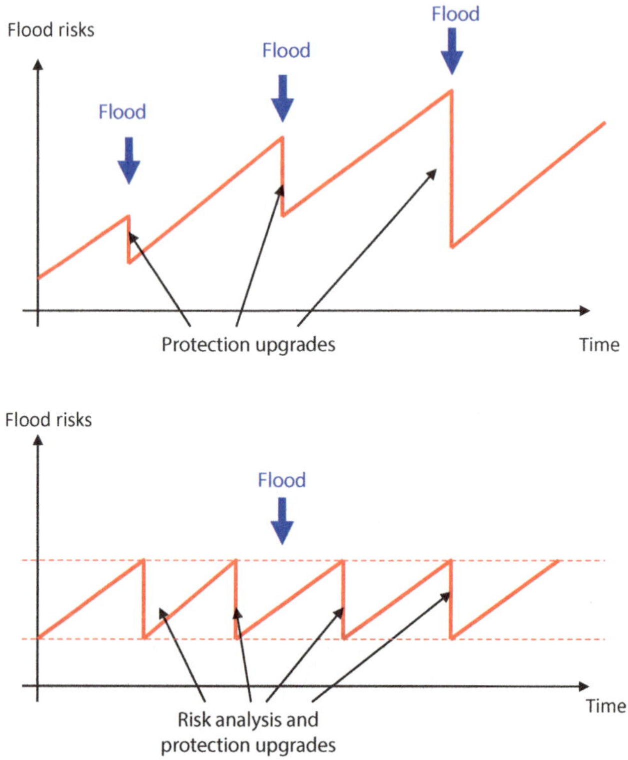


**Fig. 6.4** Two risk-management strategies. On the *top panel*, a reactive strategy, as observed in New Orleans, where flood defenses are improved after each flood. On the *bottom panel*, a proactive strategy, applied in the Netherlands, where a political process defines the maximum acceptable risk and regular risk-analysis and defense improvements make sure that the actual risk never exceeds the acceptable level. In such a situation, floods are still possible, since risk is not zero, but the risk is known and kept below a pre-determined level (Adapted from Muir-Wood and Grossi 2006)

**The main difference in the responses in New Orleans and in the Netherlands is the fact that the Netherlands created institutions to manage the risk (i.e., to measure the risk on a regular basis, to update norms and methods, and to develop response strategies), while New Orleans focused on physical protections, which were implemented in the years following each event and then neglected until a new disaster occurred.**

The consequence of flood risks is illustrated on Fig. 6.4, which shows how risk evolves over time in the two cases. In a reactive risk-management strategy, the risk is reduced thanks to protection investments only after a disaster. This approach is appropriate if the risk does not evolve over time. But if, as observed in New Orleans, the level of risk increases over time because of subsidence or economic development, then a reactive approach cannot prevent flood risks from increasing. A proactive strategy, as that of the Netherlands, reviews the risk level regularly, and implements protection investments as soon as the risk exceeds a pre-determined level. Of course, such an approach does not make disasters impossible, since risk is never canceled, but it keeps the risk within known boundaries.

**Thanks to the revision every 5 years, changes in sea level – due to subsidence or climate change – can be accounted for in risk management actions and policies. With such practices, risk management naturally contributes to climate change adaptation. This mainstreaming of climate change into disaster risk management makes it much less likely that this country will suffer large losses from climate change in spite of its vulnerability to sea level rise.**

### *6.2.2 The Adaptation Gap*

**Because natural risks are often managed in a purely empirical manner (or not managed at all), there is a potential for large gains from applying more broadly proactive risk management strategies. As shown by the case study in Mumbai (see Chap. 5), many societies are not adapted to the climate they are currently experiencing. This ill-adaptation is often referred to as the "adaptation gap", or "adaptation deficit." A climate change adaptation policy can close this adaptation gap and therefore yield short term benefits.**

There are locations where the current level of risk is clearly too high compared with the level that would be considered ideal by a risk analysis. In other locations, the level of risk might be too low, with protection costs that are larger than what would be optimal. Of course, the definition of what is optimal here is largely dependent on values and preferences, and especially "risk aversion". Two different individuals, firms, or countries may have different views on what is the optimal level of risk, some being ready to invest much more than others to reduce the level of risk.

Adaptation strategies are very different depending on whether they begin with an ideal situation where the flood risk level is at its "optimal" value, or a sub-optimal situation. This difference is illustrated in Fig. 6.5. The real situation today is that of square 1, a situation where the risk is not necessarily at its optimal value; in other terms the considered system is not adaptation to the current climate (and the difference is referred to as the "adaptation gap"). Different definitions of adaptation are represented in the table below, which is interpreted as follows:

- Going from square 1 to square 2 is the reduction of the "adaptation gap", i.e., the shift from a situation where there is no perfect adaptation to the current climate to a situation that would be optimal in the absence of climate change. This is what could be achieved by applying risk management strategies in a situation without climate change.
- Going from square 2 to square 4 is adaptation in the strict sense, i.e. the action that is necessary because of climate change alone, to go from an optimal state without climate change to a new optimal state with climate change. This type

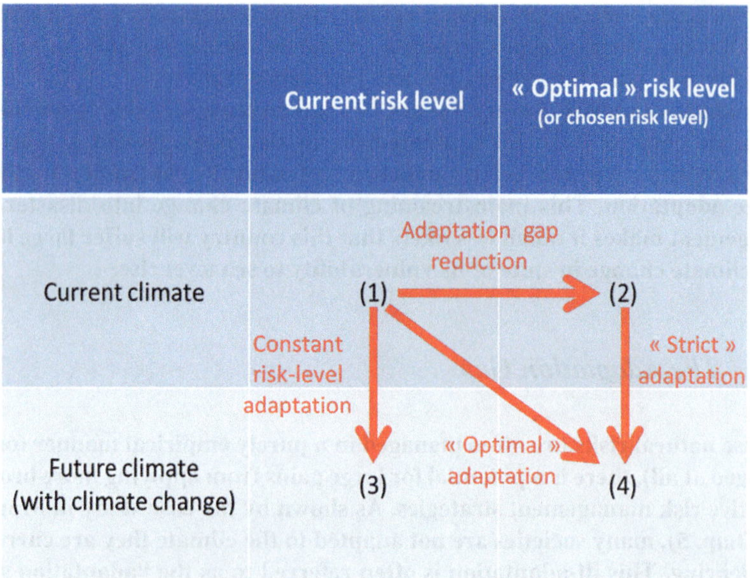


**Fig. 6.5** Illustration of the adaptation gap and of different definition of climate change adaptation

of adaptation can be qualified as "*adaptation stricto sensu*" and corresponds to actions that would not be desirable without climate change and that only become desirable because there is a change in climate.

- Going from square 1 to square 3, i.e., maintaining the risk at its initial level, can be qualified as "constant level adaptation". This type of constant level adaptation is often that which is analyzed in the scientific literature when authors begin with the premise that the current situation is optimal. Designing adaptation in a way that simply maintains the current level of risk would be sub-optimal compared with a redesign of adaptation and risk management in a consistent manner.
- The best strategy is to shift from square 1 to square 4, i.e., the passage from the current sub-optimal situation without climate change to an optimal situation with climate change. This adaptation can be qualified as "optimal adaptation," and it includes action to close the adaptation gap.

Doing so requires quite sophisticated risk and economic analysis, accounting for climate variability and climate change together, in an integrated analysis framework. One tool often used to do so is the Cost-Benefit Analysis (CBA); see Arrow et al. (1996). The next section will demonstrate the use of the CBA for risk management, and highlight the limits of this tool.[3]

[3]Another exposition is proposed in Mechler (2005).

## 6.3 Case Study: A Cost-Benefit Analysis of New Orleans Coastal Protections[4]

To illustrate the use of a CBA, this section focuses on the question of whether it is desirable to make the New Orleans flood protection system able to cope with Category 5 hurricanes. This exercise highlights the limits of a CBA analysis when dealing with a highly uncertain situation.

In a CBA framework, New Orleans would only benefit from a flood protection system able to cope with Category 5 hurricanes, compared to a system able to cope with Category 3 or 4 hurricanes, if the additional cost of the upgraded protection was lower than the expected benefits from reduced flood damages. This is not certain to be the case, as surprising as this might appear given Katrina's devastating impact, and this section will show how two evaluations can reach opposite conclusions.

To do so, we will first carry out a simple CBA using available data on the damages caused by Katrina in New Orleans. As we will see, this first CBA clearly rules out a Category 5 protection system. Then, we will show how less optimistic assumptions about anthropogenic perturbations of the environment and consideration of additional processes – namely, second-order impacts, discount rate choice, countervailing risks and side effects, risk aversion, and damage heterogeneity – can change the terms of the analysis and thus potentially justify the implementation of a Category 5 system.

### *6.3.1 A First Cost-Benefit Assessment*

To carry out a CBA of a Category 5 flood protection system in New Orleans, one needs to assess the cost, C, of such a system, and its expected benefits, B.

Assessing the cost, C, of an upgrade of the protection system is not easy, as it requires a precise definition of the system and an assessment of its construction and maintenance costs. In the very early stages of rebuilding New Orleans and its protection system, state officials estimated the cost of Category 5 protection between \$2.5 and \$32 billion. More recent and detailed estimates by Louisiana Coastal Protection and Restoration (LACPR, led by the U.S. Army Corps of Engineers) reach even larger values. In the following discussion, and for illustrative purposes, it is assumed that the cost of Category 5 protection is \$20 billion more than Category 4 protection.

**Assessing expected benefits is even more problematic, as one needs to take into account benefits of various natures (e.g., avoidance of casualties, injuries, economic losses, psychological trauma) impacting different groups of people and possibly lying far in the future.**

---

[4]This section is based on Hallegatte (2006).

Benefits can be defined as the net present value of the expected amount of damages avoided by the protection system upgrade. These benefits can, therefore, be calculated as the discounted sum, for each year from now through the lifetime of the protection system, of the annual probability that a Category 5 hurricane hits New Orleans multiplied by the difference between the damages of such a hurricane on a Category 4 versus a Category 5 protection system. This difference is discounted to take into account the fact that the same benefit is valued at a higher price when it occurs in the near future rather than further in the future.

The values of three parameters are thus necessary: the discount rate ($\delta$, see Box 6.2), the probability of occurrence ($p_n$), and the amount of avoidable damages in the year n ($d_n$). From them, expected benefits, B, are easy to calculate:

$$B = \sum_{0}^{T} p_n \left( \frac{1}{1+\delta} \right)^n d_n \tag{6.1}$$

The variable T is the lifetime of the protection, and can be assumed almost infinite, provided that adequate maintenance is provided. Assuming that damages are growing over time at the same rate as economic growth, g.

$$B = \sum_{0}^{T} p_n \left( \frac{1}{1+\delta} \right)^n d_0 (1+g)^n \tag{6.2}$$

where $d_0$ is the amount of damages a flooding of New Orleans would cause today. If the cost, C, of the flood protection system is lower than the expected benefits, B, then the system should be implemented. In spite of the difficulties already mentioned, a rough assessment of B can be made based on current information. From historical experience (i.e., by observing hurricane frequencies over the last century), one can evaluate the annual probability that a Category 5 hurricane hits New Orleans at about $p_n = 1/500$. The analysis in Chap. 5 suggests even lower probabilities.

The Office of Management and Budget (OMB), which carries out CBA of federal regulations in the U.S., uses two different discount rates to analyze policy decisions (see Appendix D, OMB Circular A-4). These two discount rates are used to assess the robustness of findings to the choice of discount rate and to capture two approaches to CBA. First, the discount rate can be calculated as the opportunity cost of capital, especially when strong capital reallocation is involved, yielding a value of 7 % in the U.S. Second, especially when the project affects consumption patterns (e.g., fiscal changes), the discount rate can be calculated as the "social rate of pure preference" used by the average American saver in his saving decisions, yielding a value of 3 %. Because the New Orleans flood protection system deals with the optimal allocation of capital, the use of the first value of 7 % can appear a priori more appropriate.

**Box 6.2: The Discount Rate**

**Cost-benefit analyses require comparing costs and benefits that occur at different points in time. This comparison is done using a discount rate, which reflects the fact that benefits and costs that occur earlier are valued more than remote ones. There are many ways of determining which discount rate should be used, and they are not consistent. Eventually, the choice of a discount rate is a political choice; but this political choice can be informed by economic theory.**

"Market" discount rates can be inferred from different market prices. For social projects, however, the use of market interest rate as discount rate can be questioned (Arrow et al. 2013). For instance, using a market rate would lead to disregarding all impacts from climate change. And market discount rates represent the preferences of private actors over relatively short time periods (up to 30 years); these preferences have no reason to be similar to society's preference, or to preferences regarding longer-term impacts.

This is why some authors have proposed to use "prescriptive" discount rate, grounded on ethical considerations and different from market discount rates. In general, a prescriptive discount rate depends on: (1) the "pure preference for the present," i.e. a term that represents pure impatience (we prefer to benefit from welfare gain earlier rather than later), (2) the economic growth that can be expected (if we are richer in the future, one more unit of consumption, e.g. one more dollar, will then be less important for our welfare) (Ramsey 1928). The discount rate thus depends on country specific factors, such as cultural and behavioral characteristics and values, and future economic growth.

**What Is Currently in Place?**

In the US, OMB guidance[5] recommends using two discount rates, equal to 7 % and 3 %. The first one is close to market rates (and the opportunity cost of capital); the second is close to what can be inferred from individuals saving decisions.

Several governments have chosen to use different discount rates for different time horizons. For instance, the UK[6] and the French[7] discount rates are decreasing over time.

---

[5]Appendix D, OMB Circular A-4.

[6]UK (2003) "Green Book, Appraisal and Evaluation in Central Government" (http://greenbook.treasury.gov.uk/).

[7]Commissariat Général au Plan, 2005. Révision du taux d'actualisation des investissements publics.

(continued)

**Box 6.2** (continued)

In France, the discount rate is at 4 % for 30 years, and then it decreases toward 2 % over time.

$$\begin{aligned} \delta_n &= 4\% && \text{if n} < 30 \text{ year,} \\ \delta_n &= \left[1.04^{30} 1.02^{(t-30)}\right]^{(1/t)} - 1 && \text{if n} > 30 \text{ years.} \end{aligned} \quad (6.3)$$

However, in developing countries, discount rates are typically higher. Discount rates in developing countries range between 8 and 15 % (ADB 2007). The difference is mainly due to resource constraints and shorter planning horizons. In a descriptive framework, there is no doubt that private and public actors' discount rates are higher in developing countries. In a prescriptive context, where the discount rate is fixed according to normative judgments, it is however questionable whether this should be the case.

**Constant or Declining Discount Rate?**

A decreasing discount rate is justified by several factors:

- **The uncertainty on future growth** (see Mankiw 1981; Gollier 2002). The discount rate applied on consumption (in dollar terms) depends on how much economic growth occurs in the future, creating an uncertainty of its value. Assuming that the uncertainty on future growth increases when looking farther in the future and that growth shocks are positively correlated over time, there is a rationale to use a discount rate that decreases over time (Gollier 2008).
- **Aggregation issues.** In an Expected Net Present Value framework, the aggregation of different individual discount rates into a social discount rate can also lead (through averaging) to a discount rate that decreases with time (Weitzman 1998). Also, the aggregation of different beliefs on what should be the correct discount rate can also lead (through averaging with Baysian probabilities) to a discount rate that decreases over time (Weitzman 2001).
- **Subsistence consumption.** If we take into account that households in some countries are very close to the subsistence level, then a prescriptive discount rate should be higher, but declining over time (as income gets larger and further from the subsistence level).
- **Equity considerations** can justify discounting differently intra-generational and inter-generational trade-offs (Stern 2006; Hallegatte 2008; Heal 2009).

The main criticism that has been made about declining discount rate is related to time inconsistency, i.e. the fact that using a declining discount rate

(continued)

**Box 6.2** (continued)
would lead to rank differently different options at different points in time. However, the reasons stated above answer this criticism: (1) if the decline in discount rate is justified by uncertainty or aggregation issues, then two decisions or ranking at two points in time are made with different information and can be different without time inconsistency; (2) if the decline is justified by the intra-generational/inter-generational difference, then two decisions at different points in time are made by different individuals and can be different without time inconsistency.

**Consumption or Utility Discount Rate?**

The discount rate applied to a given consumption impact (e.g., climate change impacts on income and consumption) or to any monetary value depends on whether this impact is correlated or anti-correlated with economic growth. If more growth leads to higher negative impacts, then the discount rate should be higher; otherwise, it should be lower (Gollier 2012). The idea is that if negative impacts occur in good states of the world, their effect on welfare is lower than if they occur in already bad states of the world (in other terms, the marginal utility of consumption is larger in bad states of the world).

An alternative to using the correlation in the discount rate is – as suggested by Arrow – to use a discount rate applied on "utility" (not on consumption), and to take into account the differences across states of the world in the assessment of climate change impacts. To do so, consumption changes need to be translated into utility changes using the "marginal utility of consumption," the value of which depends on the consumption level in the baseline scenario. Then, the discount rate is simply the rate of pure preference for the present.

Insurance and reinsurance companies (e.g., Munich Re, Swiss Re) and disaster modeling companies (e.g., RMS, EQECAT) estimate the direct damages due to any hurricane or flood, and their results are widely used as proxies for the overall economic cost of disasters. These companies estimate the cost of the New Orleans flooding at around $20 billion (RMS 2005).[8]

Taking into account direct casualties (about 1,000 people died in the flooding of the city) raises the difficult issue of attributing a cost to a loss of life. Because the expression "value of the human life" problematically suggests a market in which one

[8]Note that the losses due to the New Orleans flooding were only a fraction of the total cost of the Katrina landfall.

could buy or sell human lives, it is preferable to use the expression "incremental cost of life-saving investments" or the "amount the public is willing to devote to reducing risk in order to save an additional life" (see Box 6.3). Even though the value depends on the type of risk and the probability of occurrence of the considered event, most estimates lie between \$1 million and \$10 million in the U.S. We will use here the estimation of the U.S. EPA of \$5 million. Given this figure, the public would be willing to pay \$5 billion to reduce risks in such a way that the equivalent of the human toll of the New Orleans flood is avoided.

**Box 6.3: A Stupid Name for a Useful Concept – The Value of a Statistical Human Life**

In its Second Assessment Report, the IPCC used "values of statistical human life" differentiated by national income to assess the total economic impact of climate change, including impacts on health and mortality. Understandably, the assumption that the value of a human life depends on its income level has raised heavy criticisms on an ethical basis, and this concept disappeared in the Third Assessment Report. Recently, Ackerman et al. (2009) have repeated their rejection of this concept, viewed as highly unethical. These criticisms are understandable, and this discussion is common in disaster risk management, where human losses often represent a large fraction of total losses and have to be taken into account in the decision-making concerning risk reduction investments (e.g., dikes or early warning systems).

However, these criticisms arise mainly from the wording of this concept, the "value of a statistical human life," which uncomfortably suggests the existence of a market on which one could buy or sell human beings. Besides its wording, indeed, this concept can be useful in economic analysis and public decision-making.

For a decision-maker, indeed, there are many different ways to save lives. One can invest in the health system, for instance by financing new and better equipped hospitals or by funding additional research on frequent illness (e.g., malaria); one can invest in transport infrastructures to make them safer; one can regulate in a stronger manner local pollution in cities, thereby reducing related health effects; one can reduce speed limits on highways decreasing the frequency of deadly accident; one can introduce new building regulation to make housing less vulnerable to fire; etc. All these actions can save lives, and they cost money.

One important question is thus whether all these actions are worth being implemented and how much should be invested in each of them. In particular, it is desirable not to invest one million Euros to save a certain number of lives in one sector (e.g., the transport infrastructure) if the same million Euros could save twice as many lives in another sector (e.g., in the health sector).

(continued)

**Box 6.3** (continued)
If resources are scarce, distributing a given amount of life-saving investment among all these sectors requires measuring the relative efficiency of investments in each of these sectors.

One way of doing so would be to list all possible investments, in all sectors, to rank them accordingly to their efficiency, and to start by funding the first one, then the second one, and so on until the available resources have been spent. This method is clearly difficult to apply in practice, however, since these investments are usually proposed by many different economic actors that are difficult to coordinate, and because these investment opportunities arise at various points in time.

A more practical way is to introduce a value, which can be called the "incremental cost of life-saving investments" and is the cost of the cheapest risk-reducing investment that is able to reduce risk such that it saves – on average – one life. Assuming that all lives are equally important,[9] this value is such that all investments that can save more lives per Euro than the investment cost divided by this value should be undertaken; while all investments that save fewer lives per Euro should not be undertaken.

In this framework, this almost-value-of-a-statistical-human-life is not a measure of the absolute value of a life; it is a measure of how much it costs to save a life. In particular, this measure is not prescriptive – it does not tell how much we should spend – but it describes a given existing situation and helps in determining where it is most efficient to spend money in risk-reducing activities. This value does not measure the value of a human being, but it helps in distributing limited resources so that they save a maximum number of lives.

The fact that this value depends, in practice, on income is of particular interest in our discussion. It is a matter of fact that it costs much more to save one additional life in the US or in Europe, where health systems are well developed and risks are controlled and managed, than in many poor countries, where health systems are underdeveloped and risks undermanaged. Plenty of assessments demonstrate that limited funding could save a very large number of lives in the least developed countries, for instance through an increase in malaria drug availability, an enhancement of drinking water quality

[9]This is not always the case. In France, in the transportation sector, the government decided to invest 50 % more to avoid one death in a public-transportation accident (bus, train, etc.) than in an individual-car accident ("Instruction Cadre relative aux méthodes d'évaluation économique des grands projets d'infrastructure", 2005, available on http://www.statistiques.equipement.gouv.fr/IMG/pdf/Instruction_cadre_maj_2005_cle147216.pdf, last accessed on July 22, 2009).

(continued)

**Box 6.3** (continued)

or the provision of sanitation infrastructure. Since the cost of mitigating climate change is the same in poor and rich countries, it is understandable that reducing climate change is viewed as a less efficient way of saving lives in Africa (where other cheap alternatives exist) than in Europe (where most cheap alternatives have been implemented already).

One can wonder about the optimal value of the "incremental cost of life-saving investments." This optimal value would tell how much we should invest in life-saving investment and be used to compare life-saving investments with all other types of consumption (e.g., investing in health system vs. building a museum). When using such a trade-off, the usual criticisms to the statistical value of a human life apply, and economists have little to say to answer such a political question.

To be complete, one has also to take into account the emergency costs, such as providing health care to hundreds of thousands of people and ensuring safety and security in the affected area. This cost has been estimated at around $8 billion after Katrina.

According to these rough estimates, $30 billion seems to be a good approximation of the **direct** (as defined in Chap. 2) cost of the New Orleans flooding.[10]

Assuming that a Category 4 protection system does not reduce the damages caused by Category 5 hurricanes, which is likely since there is little difference between no levees and broken levees, the expected present benefit of a Category 5 flood protection system in New Orleans can be calculated with Eq. (6.2) at $1.3 billion with a 7 % discount rate and $6 billion with a 3 % discount rate. Both are one order of magnitude lower than the cost of building such a system. This rough estimate clearly rules out an upgrade of the protection system to make it able to cope with Category 5 storms. It might be difficult to believe that the risk of a repetition of the devastation caused by the Category 4 hurricane Katrina is not enough to justify implementation of the best possible protection system. However, our CBA suggests that it is more rational from an economic point of view to live the Katrina nightmare again in a more or less remote future.[11]

[10]In case of a repetition of Katrina's scenario, a better evacuation would probably avoid a large part of the human losses and reduce this amount of damages. It has to be mentioned, however, (i) that Katrina's track forecasts have been very good and allowed for anticipated decisions before landfall, which is not always possible, and (ii) that an evacuation is always subject to organizational problems and unexpected practical difficulties, making the human part of the damages highly variable and uncertain.

[11]An annual probability of 1/500 means that there is a 20 % chance of a Category 5 hurricane hitting New Orleans in the next 100 years, and a 33 % chance in the next 200 years.

## *6.3.2 Sensitivity Analysis*

**This estimate is, however, not very solidly grounded, as it does not take into account important processes whose impacts could be significant. To illustrate this problem, we will now review the parameters[12] of the CBA, and propose alternative estimates.**

### 6.3.2.1 Probability of Occurrence

In the first CBA, historical evidence was used to assess the probability of occurrence of a Category 5 hurricane landfall on New Orleans. This assessment cannot, however, be considered robust. Indeed, a flood protection system has a very long lifetime. Such a long lifetime arises, of course, from the long lifetime of infrastructures (dams, bridges, gates). But, above all, it comes from the fact that the flood protection system will shape the city development over an even longer time horizon. The decisions that are made on the city's protection will influence its vulnerability for at least this entire century. During this period, two mechanisms will influence hurricane risks, in addition to socioeconomic parameters like population change and asset vulnerability and value.

The first mechanism is soil subsidence, which is perturbed by human settlements and infrastructure (Hanson et al. 2011). The speed at which the soil – and therefore the levees – are subsiding is an important parameter determining the lifetime of a protection system and of the amount of potential damages.

The second mechanism is climate, which will change in the course of the century, perturbed by the increasing amount of greenhouse gases in the atmosphere. This climate change will influence hurricane risks through two channels: sea level rise and hurricane frequency and intensity. Indeed, a rising sea level makes the consequences of any storm surge more destructive, in the same way that soil subsidence does. Moreover, sea level rise and other human-induced disruptions to the Mississippi River delta (e.g., sediment deposition reduction) will worsen the floods associated with any hurricane falling on this low-lying area. Thus, the probability of floods currently caused only by Category 5 hurricanes might increase, as less powerful hurricanes could also produce such devastating floods. A second channel through which climate change modifies hurricane risks is, directly, through the probability of landfall. As discussed in Chap. 6, there is a large uncertainty on how climate change will influence hurricanes.

For illustrative purpose, we simply assume that climate change and subsidence may multiply by 5 the probability of the floods currently caused by Category 5 hurricanes, over the twenty-first century; this is in the range of possible change in

[12]Among the necessary assumptions in the CBA, it is often useful to distinguish between the political choices that must arise from a political process (e.g., discounting scheme), and the scientific uncertainties that can be – at least theoretically – solved through additional research (e.g., future probability of occurrence).

hurricane frequency (see Chap. 5). This probability would thus increase to 1-out-of-100-years in 2100. This higher probability alone would make expected benefits from protection against Category 5 hurricanes rise from $1.3 to $2.4 billion or from $6 to $23 billion, depending on the discount rate (7 % and 3 %, respectively).

One can note that with increasing landfall likelihood, the importance of the discount rate increases. This is not surprising: with increasing risks, most of the protection benefits occur in the far future, when hurricane risks are largest. Because the present value of these benefits is highly dependent on the discount rate, its importance rises.

These results suggest that climate change may have an important impact on long-term hurricane risk, even though changes in population and capital at risk will obviously be the main driver of vulnerability during the next decades.

#### 6.3.2.2 Avoidable Damages

Another major difficulty remains in the assessment of the actual damages that could be avoided through an upgrade of the protection system. Assuming that New Orleans will be reconstructed and that all displaced households will return to their original city, the damages from the Katrina landfall can be used as a proxy for the damages a future flood may cause.

As mentioned in Chap. 2, however, the direct costs, evaluated by insurance companies, may be poor proxies of overall costs, especially concerning large-scale events. According to the analyses provided in Chaps. 3 and 4, the indirect losses for the New Orleans floods are highly uncertain. In the following, given the great vulnerability of New Orleans and its vicinity (important economic activity in sensitive sectors like tourism, transportation, and energy production; low reconstruction capacity due to a large proportion of low-income population), the extent of the damages (80 % of the city under water), and the difficulties currently met in the reconstruction process, a conservative estimate of the actual overall cost of the New Orleans floods is at least double the insurers' approximation based on direct losses only; that is, $60 billion.

Using the new values of event probability and potential damages, the expected benefit of an upgraded protection system would be $4.8 billion with a 7 % discount rate and $46 billion with a 3 % discount rate.

#### 6.3.2.3 Countervailing Risks and Other Side Effects

**It is also necessary to take into account the possible side effects implied by the implementation of a large-scale protection system. These side effects can yield ancillary benefits like infrastructure improvement, as mentioned by Allenby and Fink** (2005)**, or create or increase other risks, referred to as countervailing risks by Wiener** (1998)**, who calls for a broader accounting of them in risk management.** The importance of these side effects will be heavily dependent on the design and practical implementation of the protection system. In particular, huge negative consequences would certainly result from the implementation of an ambitious flood protection system that is not carefully maintained over the long term. In this worst case scenario, the existence of the protection system would raise investment and population in the so-called protected area, which would not be protected any more after a few decades of negligence, making vulnerability even larger than if no protection was implemented in the first place. As a consequence, the implementation of a protection system must be considered a long-term commitment.

Also, avoiding negative outcomes from the future flood protection system requires careful design and implementation, in order to protect already urbanized areas without steering additional urbanization toward unprotected flood-prone locations. In this respect, the future flood protection system in New Orleans is certainly not only a system of dams, bridges, and gates. As shown in Sect. 6.1, risk management requires a policy mix, not a single action or measure. In the New Orleans case, risk management should also include an important set of new regulations for future urban development. A wisely designed flood protection system should protect selected areas with dams and levees, and ensure, through land-use regulations, that investments are not attracted to unprotected areas. Hopefully, increased experience with flood management and the high visibility of the project will foster a flood protection plan that limits the negative effects and promotes positive ones, making the overall consequences of these side effects positive or, if this proves impossible, negligible compared with direct costs and benefits.

#### 6.3.2.4 Choice of the Discount Rate

As already mentioned, the CBA of a flood protection system has to deal with very long time horizons, making the value of the discount rate controversial.

Clearly, as illustrated by our comparison of 3 and 7 % discount rates, the influence of this political choice is large. In other terms, decision making concerning protections against very infrequent events is highly dependent on the weight attributed to the well-being of future generations.

#### 6.3.2.5 Risk Aversion

**A society that would use the previous method to assess a protection system is called risk-neutral. A risk-neutral agent is indifferent to risk; i.e., it does not see any difference between losing \$1 with certainty and having a 10 % chance of losing \$10, because the expected loss is the same in both cases. Theoretically, such an agent would never pay for insurance. Regarding protection against large-scale floods, however, there are good reasons to justify risk-averse behavior: people might indeed prefer to pay an additional amount of money (a risk premium) to avoid the risk of costly and deadly floods (see also Box 2.6).**

To take this effect into account, the CBA needs to be carried out taking into account this risk aversion. It is possible to do so by doing the CBA using a social utility function, which relates total income (or consumption, $C$) to the social welfare u that can be derived from it: $u = u(C)$. If the derivative of $u$ with respect to $C$ is not constant (i.e. if an additional $1 of consumption yields a different amount of welfare, depending on the initial level of consumption), then there is a risk aversion.

**A utility function with risk aversion assumes that the increase in utility due to a $1 gain is smaller – in absolute value – than the decrease in utility due to a $1 loss. As a consequence, the risk of gaining or losing $1 with equal probability lowers the expected utility and is, therefore, equivalent to a certain financial loss, which is referred to as the risk premium or the equivalent-certain outcome.**

Risk aversion is defined using the Arrow-Pratt measure of absolute risk aversion (ARA):

$$ARA = -\frac{u''(C)}{u'(C)}, \tag{6.4}$$

or the relative risk aversion (RRA):

$$RRA = -\frac{Cu''(C)}{u'(C)} \tag{6.5}$$

To do the analysis in that case, one compares the net present utility $U_b$ in a baseline scenario with no protection against a category 5 hurricane and the net present utility $U_p$ in a scenario with such a protection. In this scenario with protection, the consumption in year 0 is reduced by the initial investment cost in the protection (i.e. by $C$, or $20 billion), but the reduction in consumption due to a category 5 hurricane only appears in the scenario without protection.

$$U_b = \sum_{n=0}^{n=\infty} \left(\frac{1}{1+\rho}\right)^n \left[P_n u\left((C_0 - d_0)(1+g)^n\right) + (1-p_n)\, u\left(C_0(1+g)^n\right)\right] \tag{6.6}$$

$$U_p = u(C_0 - C) + \sum_{n=1}^{n=\infty} \left(\frac{1}{1+\rho}\right)^n u\left(C_0(1+g)^n\right) \tag{6.7}$$

**The protection is implemented if the protection increases the utility, i.e. if $U_b < U_p$.**

**Note that the discounting of utility should not be done with the same discount rate as for consumption fluxes (see Box** 6.2). In the latter case, the discount rate includes the fact that the utility provided by $1 of additional consumption decreases with wealth. When discounting utility, this effect is already taken into account in the (convex) form of the utility function. As a consequence, utility fluxes are discounted with the pure preference for the present (ρ) only. Assuming a 2 % growth rate and an elasticity of the marginal utility ν equal to one, the equivalents of the 3 % and 7 % discount rates suggested by the OMB are 1 and 5 % pure preference for the present.

**However, if we assume that the damages due to a hurricane landfall are perfectly shared among the whole population of the U.S., then risk aversion can be neglected.** In that case, indeed, the damage per capita is small (a few hundred US$ per capita), i.e. $d_0$ is small compared with $C_0$. In the same way, the protection cost $C$ remains small compared with total consumption $C_0$. In this case, the equations can be linearized and taking into account risk aversion only adds a few million dollars to protection benefits, for all reasonable values of the risk aversion.

**In such a situation, the Arrow-Lind theorem (Arrow and Lind** 1970) **demonstrates formally why risk-aversion can be neglected, supporting the choice of the States that consider self-insurance as a basic principle (e.g., France). Indeed, using a utility function with a constant relative risk aversion of one, as suggested in Arrow** (1974) **for households in developed countries, the risk-premium is negligible.**

#### 6.3.2.6 Damage Heterogeneity

**The picture is different, however, if a substantial part of the damages impacts only a small fraction of the population. Indeed, when the utility function is not linear, it is not equivalent for a group of 10 people either to lose $1 each, or to know that one of them will lose $10.**

The consequences of these factors on the CBA analysis of a hurricane landfall can be very significant. In this case, indeed, individual losses become significant (as large as 50 % of annual consumption for the affected population) and risk aversion appears far from negligible. Here, as an illustration, we assume that one third of the cost of a category 5 hurricane is shared among the entire U.S., while two third of the losses are shared only by the 3 million affected inhabitants. In the case of Katrina, a large fraction of the losses have been shared at the federal level, through tax-paid support to the affected population; but a large fraction of the losses is not transferable, from psychological trauma and loss of friends or relatives to the need to rebuild a home.

**In that case, the utility cannot be calculated at the aggregated level, using the country consumption level, like in Eqs.** (6.6 **and** 6.7). **It is instead necessary to calculate the utility for each inhabitant, and to aggregate them to obtain a "social utility." There is no consensual way of aggregating the utilities of different individuals, and there is no way of observing utility functions (see Box 6.4). Here, we will use the simplest way, which is to assume that all individuals have the same utility function, and to sum the individual utilities to calculate the social utility.**

For the population that is potentially affected by the hurricane, the baseline and protection utilities ($U_b^a$ and $U_p^a$ are given by:

$$U_p^a = \sum_{n=1}^{n=\infty} \left(\frac{1}{1+\rho}\right)^n \times \left[ P_n u \left( \left( c_0 - \frac{(1-\alpha)\, d_0}{P^a} \right) (1+g)^n \right) + (1-p_n)\, u\left(c_0(1+g)^n\right) \right] \tag{6.8}$$

$$U_p^a = u\left(c_0 - \frac{C}{P^t}\right) + \sum_{n=1}^{n=\infty} \left(\frac{1}{1+\rho}\right)^n u\left(c_0(1+g)^n\right) \tag{6.9}$$

where α is the share of total losses that is transferred to the non-affected population, $c_0$ is the individual consumption level, $P^t$ is the total population in the country and $P^a$ the population potentially affected by the hurricane. Losses from the hurricane are shared among the affected population $P^a$, while the protection costs are shared among the total population $P^t$.

For the population that is not affected by the hurricane, the baseline and protection utilities ($U_b^n$ and $U_p^n$ are given by:

$$U_b^a = \sum_{n=0}^{n=\infty} \left(\frac{1}{1+\rho}\right)^n \left[ p_n u \left( \left( c_0 - \frac{\alpha d_0}{P^t} \right) (1+g)^n \right) + (1-p_n)\, u\left(c_0(1+g)^n\right) \right] \tag{6.10}$$

$$U_p^a = u\left(c_0 - \frac{C}{P^t}\right) + \sum_{n=1}^{n=\infty} \left(\frac{1}{1+\rho}\right)^n u\left(c_0(1+g)^n\right) \tag{6.11}$$

The total utility in the baseline and protection scenarios are $U_b$ and $U_p$:

$$U_b = P^a U_b^a + \left(P^t - P^a\right) U_b^t \tag{6.12a}$$

$$U_p = P^a U_p^a + \left(P^t - P^a\right) U_p^t \tag{6.12b}$$

In the following, we use a logarithm utility function that links individual consumption with individual utility:

$$\mathrm{U} = \ln\ (\mathrm{c}) \tag{6.13}$$

Applied with our parameters, assuming that losses from a category 5 hurricane are equal to \$60 billion, that probability of occurrence will increase from 1 in 500 years now to 1 in 100 years in 2100, this approach makes the protection worth building if its cost is lower than \$65 billion with a 1 % pure preference for the present or 3 % discount rate (instead of \$46 billion with homogeneous losses) and \$6.7 billion with a 5 % pure preference for the present or 7 % discount rate (instead of \$4.8 billion).

These results should be taken with care, because they are highly dependent on assumptions on the factor $\left(c_0 - \frac{\alpha d_0}{P^t}\right)$, i.e. share of losses that cannot be shared with the rest of the population. In this framework, if this factor reaches zero, i.e. if the disaster is strong enough to bring remaining consumption to zero, then the protection is worth being built at any price. This is because utility tends to minus infinity when consumption tends to zero, assuming that zero consumption is impossible or simply unacceptable.

**This sensitivity illustrates the lack of robustness of the CBA when costs and benefits are not small marginal changes but imply large changes in consumption or even non-monetary losses such as fatalities and casualties.**

To represent the role of basic needs like access to food, water and housing, it is also possible to use a utility function with which utility goes to minus infinity for nonzero values of consumption. In that case, the utility can be expressed as:

$$U = \ln\left(c - c_{bn}\right),$$

where $c_{bn}$ is the level of individual consumption that corresponds to basic needs.

**In this framework, if risk aversion is non-zero, utility losses are increasing nonlinearly with the amplitude of individual losses. It means that any measure that leads to a larger sharing of the losses without changing the amount of losses would lead to an increase in utility. It means that solidarity or insurance is always utility increasing.**

This conclusion is valid when the risk level is fixed and independent of individual and collective choices (like for the risk of a meteorite impacting earth). It needs to be nuanced if risk-taking is affected by choices. In that case, it is necessary to take into account moral hazard, i.e. the fact that the potentially affected population transfers away one fraction of its losses and may be tempted to take more risk. A flood protection system funded through nationwide taxes, like a uniform insurance premium, can indeed constitute an incentive for people to settle in at-risk areas, as they do not pay for the full risk their location choice creates.

**Box 6.4: Utilitarism and Decision-Making**

The use of social utility functions, often referred to as "utilitarism", is based on the work of Bentham (1789), Marshall (1890), and Pigou (1920). This approach has however been criticized because it is classically based on the sum of utilities only, therefore neglecting all distributive issues. If a policy increases the utility of an individual A and decreases the utility of an individual B, it should be implemented if the increase is larger than the decrease.

Rawls (1971), for instance, criticized utilitarism on an ethical basis and proposed the idea that inequality is justified only if it benefits the worst-offs

(continued)

**Box 6.4** (continued)
in a population. It means that a policy is justified only if it benefits the poorest. Of course, this approach is not meant to be applied too strictly, neglecting all policy consequences except the ones on the poorest. But this approach calls for a specific analysis of policy impacts on the poorest, and for attributing a high weight to these impacts in the decision-making process.

Robbins (1938) made an even stronger criticism of utilitarism by questioning the possibility of interpersonal utility comparison. If utilities are not observable or comparable, indeed, one cannot assess a policy that increases the utility of some individuals and decreases the utility of others.

Alternatives to utilitarism have their own issues. Arrow's (1950, 1951) "impossibility theorem" shows that all voting procedures – except dictatorship – lead to inconsistencies in choices.

#### 6.3.2.7 Taking into Account Pre-existing Inequality

Equations (6.8, 6.9, 6.10, and 6.11) assume that pre-disaster individual consumption is identical for the affected and non-affected population. In practice, low-income people are more likely to belong to the affected population. The method presented in the previous section can easily accommodate for differences in income, by replacing $c_0$ by differentiated consumptions for the affected and non-affected population.

**In general, introducing income differences with non-zero risk aversion gives more weight to the impacts on the poorer population and less to impacts on the wealthy.** In this framework, it is worth making a financial transfer from the wealthier to the poor, even if this transfer is costly (i.e. some efficiency can be lost in exchange for more equity). Giving more weight to the poor in a decision-making methodology is also referred to as using "distributional weighted" in Cost-Benefit Analysis, and it can be done using a nonlinear utility function – as we did in the previous section – or simply attributing a higher weight to the costs and benefits of the poorest than to that of the richest (Harberger 1978, 1984).

This type of calculation requires a note of caution, however. It attributes a higher social value to the options that transfers income from the rich to the poor. In the case of natural disasters protection, for instance, it attributes a larger value to an investment that is less efficient, but that protects a poor population. But one has to consider also the possibility of combining a risk reduction investment based on an efficiency criterion only (i.e. considering the aggregated consumption as in Eqs. (6.6 and 6.7) instead of aggregated utility as in Eqs. (6.8, 6.9, 6.10, and 6.11)), with a distributional policy that could realize an equivalent transfer to the poor at a lower cost.

**If one assumes that transfers can be done at zero cost (through financial transfers that do not impact behaviors), then the efficiency and the equity issues can be separated: an investment should be made on the basis on an efficiency criterion only, and any detrimental distributional impact can be corrected ex**

**post with transfers.** In the real world where transfers are always costly, the question is whether there is a more efficient way to transfer resources to the poor; it might not be optimal to "use" disaster risk reduction to tackle distributional issues.

### *6.3.3 Cost-Benefit Analysis Under Uncertainty*

**When there is uncertainty on the value of one or several parameters, as in our illustrative example, it is possible to represent this uncertainty as a set of possible "states of the world", where each state of the world corresponds to a set of parameter values. In such a situation, it is sometimes possible to attribute "subjective probabilities" to the different possible states of the worlds, and to evaluate the expected benefits as the probability-weighted average of the benefits in the different possible states of the world.**

Coming back to our simple CBA, the benefit of such a protection is given by:

$$B = \sum_{0}^{T} p_n \left(\frac{1}{1+\delta}\right)^n d_0(1+g)^n, \tag{6.14}$$

and this benefit clearly depends on the value of the probability of occurrence $p_n$. As already stated, if the probability increases to 1-out-of-100-years in 2100 instead of staying constant, the expected benefits from protection against Category 5 hurricanes rise from \$1.3 to \$2.4 billion or from \$6 to \$23 billion, depending on the discount rate (7 % and 3 %, respectively).

If we assume that there is a P = 1/3 probability that the category-5 probability stays constant, and a (1 – P) = 2/3 probability that it increases up to 1-out-of-100-years in 2100, then the expected benefit can be written:

$$B = P\sum_{0}^{T} p_n \left(\frac{1}{1+\delta}\right)^n d_0(1+g)^n + (1-P)\sum_{0}^{T} p_n' \left(\frac{1}{1+\delta}\right)^n d_0(1+g)^n, \tag{6.15}$$

where $p_n$ stays constant and $p_n'$ increases up to 1-out-of-100-years. In that case, the expected benefit is equal to: 1/3*1.3 + 2/3*2.4 = \$2 billion with a discount rate of 7 %, and 1/3*6 + 2/3*23 = \$17.3 billion with a discount rate of 3 %.

**This classical method is widely used in situations of uncertainty. But these subjective probabilities for possible future states of the world should not be confused with objective probabilities, based on frequency**. When one says that a dice has a 1/6 probability to give a 6, it means that if one rolls the dice an ever increasing number of times, the number of 6s divided by the total number of draws will tend to 1/6. This probability is thus a way to describe a frequency. With climate change and subjective probabilities, the situation is different because there can be only one draw, and there is only one good answer: climate change will increase hurricane intensity or not. **In that case, assuming that there is a given probability that climate change will increase hurricane intensity only describes our belief in the effect of climate change, but it cannot be related to a frequency.**

**The problem is that – for climate change – we do not have a strong methodology to assess these subjective probabilities.** They cannot be fully based on the past, because climate change is a new process for which we have no equivalent in the past. Some have proposed to use various models to assess the uncertainty: if 50 % of the models predict an increase in hurricane intensity then we could attribute a 50 % likelihood to such an increase. This method is inappropriate, however, because there is no reason why existing models should represent the real uncertainty: they may all have the same flaw, since they are all based on the same incomplete and imperfect knowledge.

The case of sea level rise is telling. In the 2007 IPCC report, sea level rise in 2100 was projected between 18 and 59 cm, depending on models and emission scenarios. But all models used to make this assessment did not include properly ice dynamics, especially for the Greenland ice sheet. Since ice dynamics has been shown to play an important role, sea level rise projections had to be revised leading to an increase in "perceived uncertainty". IPCC (2013) increased the upper ranges of sea level rise projections for 2100 at more than 90 cm. Subjective probabilities based on the 2007 models would have produced a highly biased estimate of sea level rise probabilities, and thus highly biased decisions.

## 6.4 Case Study: Early Warning Systems in Developing Countries[13]

This section proposes an assessment of the costs and benefits of investing in the production of hydro-meteorological information for disaster early warning and evacuation systems. Because the analysis is global, it cannot be based on a precise assessment of the losses that can be avoided in each country, as a function of the type of early warning that is available. Considering the difference in socio-economic context and in environmental conditions, doing so would be impossible. As a result, we have to rely on a simple analysis. Here we assess the benefits from early warning systems where they are the most developed and efficient (using Europe as a case study), and we use simple approximations to estimate how these benefits could be transferred in developing countries where hydro-meteorological and warning systems are less developed. We are therefore not providing a precise estimate of the economic benefits, but only orders of magnitude. But as we will see, such orders of magnitude are sometimes sufficient to make a decision.

[13]This section has been published as Hallegatte (2012) and has been written for the 2012 Copenhagen Consensus.

The analysis starts by investigating benefits from early warning systems in Europe, in terms of saved lives and reduced disaster asset losses. It then uses this evaluation to estimate the potential benefits of providing similar services in developing countries (Sect. 6.4.1). It then assesses the other economic benefits that could be derived from the same hydro-meteorological information that is needed for early warning (Sect. 6.4.2). These benefits are linked to a better optimization of economic production, and can be estimated in Europe to serve as the basis for an estimate for developing countries. Section 6.4.3 assesses the cost of providing this information, and Sect. 6.4.4 concludes with a cost-benefit analysis of doing so.

## *6.4.1 Benefits from Early Warning and Preparation Measures*

Weather forecasts enable the anticipation of, and preparation for, extreme events like heat waves, cold spells, windstorms, thunderstorms and floods. Corresponding benefits accrue in two broad categories: the protection of persons and assets (prevention) and emergency preparation. Early warnings enable the protection of persons in many ways. Individuals can for instance avoid road trips when floods are forecasted, they can move vehicles out of flood zones and they can implement mitigation actions (e.g., sandbagging). Organizations and businesses can do the same. For instance, schools and businesses can be closed to avoid unnecessary trips and risks. In case of intense events, evacuation is also possible. Asset protection can also offer large benefits. Preparing a house before a hurricane (e.g. by covering windows) can reduce damage by up to 50 % (Williams 2002). A study in Germany (Merz et al. 2004) shows that in the residential sector, one-third of the damage concerns the non-fixed contents, i.e. the house contents that could be saved, thanks to an early warning, by being moved out of vulnerable places (e.g. moved to the second floor). This proportion is only 10 % in the infrastructure sector, but it grows to 60 % in services and 80 % in the manufacturing sector.

### 6.4.1.1 Illustration on Europe

Asset Losses

Thieken et al. (2007) and Kreibich et al. (2005) report on the Elbe and Danube floods in 2002. They show that 31 % of the population of flooded areas implemented preventive measures. These measures include moving goods to the second floor of buildings (applied by more than 50 % of the inhabitants who implemented prevention measures), moving vehicles outside the flood zone (more than 40 %), protecting important documents and valuables (more than 30 %), disconnecting electricity and gas supplies and unplugging electric appliances (more than 25 %) and installation of water pumps (between 2 and 10 %).

Among the inhabitants who did not implement any measure, 65 % said that they had been informed too late, and about 20 % said that they were not at home and could not do anything. For this population, it seems that an earlier warning would have allowed a better preparation and lower subsequent damage.

There is also a large potential for businesses. The International Commission for the Protection of the Rhine (2002) estimated that 50–75 % of flood losses could be avoided thanks to emergency preparation measures. For instance, moving toxic materials and chemicals to safe places prevents local pollution (such as observed after hurricane Katrina flooded New Orleans). Machines and equipment can also be moved to avoid damage. For instance, large savings are possible in the transport sector, through moving transport equipments (trains, buses, etc.) out of dangerous areas. In addition, anticipating transport perturbations reduces the costs and complications of managing passengers blocked during their journey.

However, this potential gain is not easy to capture, especially because of the difficulties associated with disseminating the warning: according to Kreibich et al. (2007), 45 % of businesses did not receive the warning directly from the authorities before the 2002 flood in Saxony, even though this warning was transmitted 20 h before the flood.

In spite of this problem, almost 70 % of all businesses implemented emergency prevention measures, often thanks to informal contacts that helped disseminate the warning. According to the study, 7 % of all equipment was fully protected, and 75 % partially protected, thanks to prevention measures. Concerning inventories and production, 10 % could be fully protected, and 70 % partially protected.

Studies show that the warning timing was critical: businesses that protected their equipment or inventories were those that received the warning early enough. This is confirmed by earlier findings by Day (1970), who found that avoided losses grow to near 35 % when lead time exceeds 36 h. According to Carsell et al. (2004), a warning emitted 48 h before a flood enables the overall damage to be reduced by more than 50 %.

Barredo (2009) finds that floods cost on average four billion euros per year Europe (EU27) (normalized costs calculated over the 1970–2006 period). Assuming that the warning reduces losses by 10 %, and if only half of the floods are forecasted, the benefits from early warnings could reach 200 million euros per year. Using Carsell's estimate, and assuming that 75 % of the floods can be forecasted, the benefits would reach 1,500 million euros per year.[14]

According to Swiss Re (2006), storms have cost on average about 2.6 billion euros per year in Europe. If weather forecasts help reduce these losses by 10 % or by 50 % – thanks to the same types of actions than before a flood – the corresponding gains lie between 260 and 1,200 million euros per year. For floods and storms, the total could thus lie between 460 million and 2.7 billion per year. These gains represent between 0.003 and 0.017 % of European GDP.

[14]We assume here that false alarms have no cost, which is not the case, especially in case of large scale evacuations.

Human Losses

It is more difficult to produce an economic estimate for individual safety. There are many weather-related threats to safety in Europe, such as floods (e.g., le Gars in France in 2002; Lower Silesia in 2005), winter storms (e.g., 1999 storms Lothar and Martin in France; Kyrill in France and Germany in 2011), heat waves (e.g., the 2003 summer) and cold spells (e.g., 1984–1985 in France; 2001 in Hungary; 2010 in Poland), and avalanches (e.g., in les Orres in France in 1998).

In Europe, severe winter-storms have a return period of about 10 years and often lead to dozens of casualties. The 2003 heat wave caused about 70,000 deaths in Europe, and the 2006 one led to about 2,000 deaths. Avalanches cause on average 32 deaths per year in France, for 4.5 million ski tourists.[15]

It is difficult to assess how many lives prevention and early warning save each year, even though local actors consider these tools critical to ensure population safety. Examples of countries where early warning is considered seriously and where prevention emergency actions are well organized (e.g., Cuba) show that casualties can approach zero, apart from really exceptional events.

Other scenarios also need to be considered, such as technological catastrophes (accidents in a chemical plant or a nuclear plant). In these cases, the capacity to forecast winds, and thus the trajectory of the contamination cloud, can easily save hundreds of lives. The need to predict the trajectory of radioactive leaks was recently illustrated in Japan by the Fukushima nuclear accident. Even though the likelihood of using this capacity is fortunately very small, the damage avoided can be so large that this possibility needs to be accounted for in the social value of forecast capabilities.

Even in non-extreme situations, hydro-meteorological information plays a large safety role in many outdoor activities (e.g. sailing, hiking, skiing). Specialized services help thousands of people avoid being surprised at sea by a storm. These services have a large audience and probably help avoid hundreds of accidents each year.

Even more important, this information is used for maritime and air traffic.[16] Indeed, these operations depend on detailed weather information and, in the absence of meteorological information, it is unlikely that passenger air travel would be safe enough to be commercially viable in its current form.

Another component of forecast value is the ability to prepare emergency services before an event occurs. During the few hours before an intense weather event much can be done to increase the efficiency of emergency services. In 2002 during the floods in the Gard, for instance, 22 out of the 26 French helicopters able to conduct rescue airlifts were pre-positioned in the flood area, thanks to the

[15]Source: Dossier d'information « Avalanche » du Ministère de l'Aménagement du Territoire et de l'Environnement, 2000, available on http://www.prim.net.fr.

[16]According to the « Bureau Enquête Accident », 7.5 % of plane accidents have meteorological causes (http://www.bea-fr.org/etudes/stat9798/stats1997-1998.htm)

forecasts. According to local emergency services, this pre-positioning saved about one hundred lives, compared with a situation in which it would have taken hours to move helicopters to the affected areas.

It is important to stress that the benefits from forecasts depend largely, and nonlinearly, on their accuracy and on the trust of the population. Also, threshold effects are important. For instance, if it were possible to predict exactly flash-floods, including their location, it would be possible to evacuate the at-risk areas, and reduce human losses to zero without any expensive investment in flood protections. But the decision to evacuate cannot be made if the probability of false alarm is too high (or if the warning area is too large): after a few unnecessary evacuations, the trust in the warning system is likely to disappear, and the warning system becomes useless. This problem is illustrated by the case of New Orleans, which had been unnecessarily evacuated twice (for hurricane George in 1998 and Ivan in 2004), making it more difficult to convince inhabitants to leave before hurricane Katrina. If the risk of false alarm becomes low enough to create and maintain trust and allow for significant prevention measures before disasters, a limited improvement in forecast accuracy can thus lead to a large increase in societal benefits. Trust should also be built through openness and communication from the specialists to the public, to explain the limits of forecasts and warning.

Taking into account these numbers, one can assume that hydro-meteorological information saves at least 200 lives per year in Europe, which is an extremely conservative estimate. A more likely estimate is about 800 lives per year.

Using standard values of the "statistical value of a human life" from the Boiteux Report on the transport sector in France,[17] i.e. 1 million euros per life, the corresponding benefits can be estimated at between 200 and 800 million euros per year.

#### 6.4.1.2 Potential Impact in Developing Countries

**Adding up asset and human losses leads to an estimate of the annual benefits from early warning in Europe between 660 million and 3.5 billion euros per year.**

These estimates are useful for our analysis because they provide an illustration of how much can be gained at the global level with current technologies and state-of-the-art modeling and observation systems. Of course, some of the potential benefits from early warning are already realized in the world (especially in developed countries such as Europe, the US, Japan, etc.). The questions are thus (i) How much of these benefits are already captured at global scale? (ii) How much would it cost to capture the full benefit potential?

[17]Instruction Cadre relative aux méthodes d'évaluation économique des grands projets d'infrastructure, 2005, http://www.statistiques.equipement.gouv.fr/IMG/pdf/Instruction_cadre_maj_2005_cle147216.pdf

It is difficult to estimate the existing availability of early warning. Some countries have introduced efficient systems in recent decades (e.g., Bangladesh and its hurricane early warning system). But in other places, basic observation systems are simply not in place to allow for implementing early warning.

Several institutions (including WMO, Golnaraghi 2012) review the ability of national hydro and meteorological services to fulfill their missions. Subbiah et al. (2008) distinguish four groups of countries. Group 1 includes countries with no basic hydro-meteorological services, where the benefits are likely to be close to zero; we assume that 10 % only of potential benefits are already realized in these countries. Group 2 includes countries where hydromet services exist but are not fully operational; we will assume that these countries realize only 20 % of the benefits achieved in Europe. Group 3 includes countries with well functioning hydromet services but with gaps in the chain from data production to early warning systems; there, we assume that only 50 % of the European benefits are achieved. The last group includes countries where hydromet services and early warning systems are comparable to European ones, and where 100 % of the European-type benefits are achieved.

There is no automatic relationship between these groups and country incomes, but richer countries have better hydromet services and early warning systems. As a consequence, we will assume that Groups 1 to 4 can be mapped to low income, low middle, high middle, and high income countries.

### Avoided Asset Losses

Focusing first on asset losses, we assume that all countries face the same level of risk, and that early warning can provide the same relative benefits in terms of avoided disaster losses. In other terms, a European-like early warning system would reduce disaster-related asset losses in all countries by between 0.003 and 0.017 % of their GDP. This is conservative, for instance, because developing countries are affected by different hazards than industrialized countries (e.g., tropical storms).

The share of losses actually avoided is assumed to depend on the type of country: 10 % of these benefits in low-income countries; 20 % in lower middle income, 50 % in upper middle income, and 100 % in high income countries. The difference between the European potential and the share actually avoided provides an estimate of the additional benefits that could be achieved if hydro-meteorological services were upgraded to European standards (see Table 6.1).

**This estimation suggests thus that generalizing the quality of hydromet services and early warning systems that can be found in developed countries could yield benefits *in terms of avoided asset losses* between 300 million and 2 billion USD per year in developing countries, thanks to lower disaster losses.**

These values are supposed to increase like economic growth even though, as discussed in Chap. 5, disaster losses growth can be slower or quicker than economic growth. This assessment is probably an underestimation because it does not take into account indirect losses from natural disasters or the possibility of poverty traps

**Table 6.1** Potential benefits from avoided asset losses thanks to early warning (with European-standard hydro-meteorological services), and share of these benefits actually realized with current services

| | GDP (million USD) | Potential (European-like) benefits | | Ratio of current vs potential benefits | Estimation of actual benefits | | Benefits from improved services | |
|---|---|---|---|---|---|---|---|---|
| | | Low estimate | Likely estimate | | Low estimate | Likely estimate | Low estimate | Likely estimate |
| Low income | 413,000 | 12 | 69 | 10 % | 1 | 7 | 11 | 62 |
| Lower middle income | 4,300,000 | 122 | 714 | 20 % | 24 | 143 | 97 | 572 |
| Upper middle income | 15,300,000 | 433 | 2,542 | 50 % | 217 | 1,271 | 217 | 1,271 |
| High income | 43,000,000 | 1,217 | 7,145 | 100 % | 1,217 | 7,145 | – | – |
| TOTAL | 63,013,000 | 1,784 | 10,470 | | 1,459 | 8,565 | **324** | **1,904** |

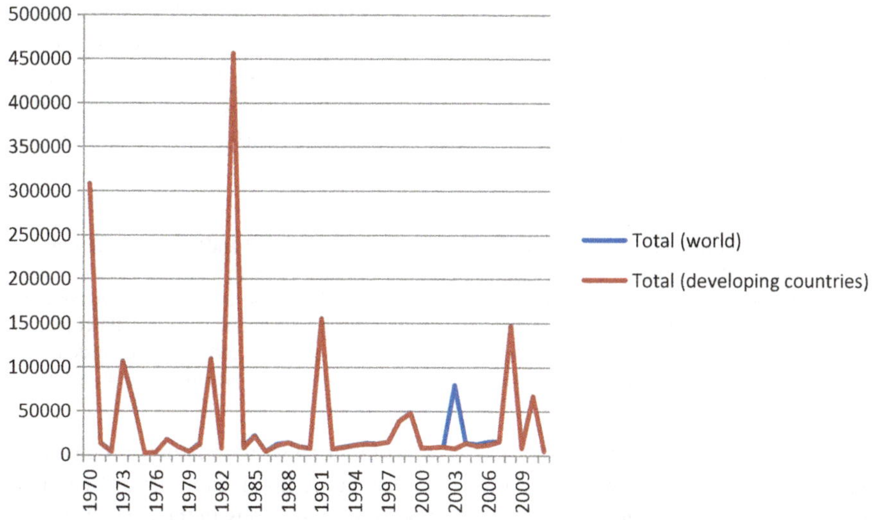


**Fig. 6.6** Number of people reported killed by weather-related natural disasters (1975–2011), in developing countries and at the world level. There is no significant trend in these series (Data from EM-DAT: The OFDA/CRED International Disaster Database)

at the household, regional, or even national level (see Chap. 2 of this book). This assessment does not account for the possibility that reduced disaster losses can lead to accelerated economic growth.

### Human Losses

In terms of life saved, the calculation is more difficult, because the lower quality of housing and infrastructure (including coastal and river flood protections) in developing countries means that the role of early warning is even more important than in developed countries.

According to the OFCA/CRED EM-DAT database, weather related extreme events killed on average 43,000 persons per year in developing countries between 1970 and 2011, and there is no visible trend in this figure (see Fig. 6.6). Since the total population in 2011 in developing countries was approximately 5.7 billion persons, there is an annual death probability of 7.5 per million due to weather events.

In developed countries, the death toll is 2,500 persons per year (mostly from heat waves). The total population in developed countries is approximately 1.1 billion persons, and the annual death probability is 2.2 per million inhabitants. The ratio of the death probability in developing countries to the death probability in developed countries is thus equal to 3.4.

This difference is not due only to early warning: housing and infrastructure quality, disaster protection (dikes, drainage systems, etc.) are also important. Differences in climate (e.g., exposure to tropical storms) also play a role. But early warning plays a huge role, as is illustrated by the case of Bangladesh and coastal floods. Paul (2009) reports on how hurricane Sidr caused many fewer deaths (about 3,400 deaths in 2007) than other similar storms (such that Gorky that killed more than 140,000 people in 1991). The main explanation is the good forecast of the storm, which allowed warnings to be issued early, and the early warning and shelter systems.

According to Paul (2009):

> The Cyclone Preparedness Program (CCP) was established in 1972 and was charged with developing effective cyclone preparedness measures for residents of the Bangladesh coastal areas (Khan and Rahman, 2007). It was jointly set up by the Bangladesh Ministry of Disaster Management and Relief (MDMR) and the Bangladesh Red Crescent Society (BDRCS). CPP activities are centered around three infrastructures: a cyclone early warning system, public cyclone shelters for pre-disaster evacuation, and shelters to provide protection for cattle during storm surges.

The CCP has been reinforced since cyclone Gorky in 1991, with better forecast and warning capacities, and twice as many volunteers being involved in the diffusion of warnings and evacuation orders. There is little high-technology in the early warning system:

> Many Red Crescent volunteers, local government officials, workers from NGOs, and some villagers immediately joined the CCP volunteers. All traveled to threatened communities and disseminated cyclone warning and evacuation orders via megaphones, handheld bull-horns, bicycle-mounted loudspeakers, and house-to-house contacts. They also evacuated people in the path of the cyclone into cyclone shelters. In addition, fishing boats and trawlers over the North Bay of Bengal were instructed to immediately return to inland river ports.

This simple low-technology system saved many lives, in spite of the lack of maintenance of shelters, with only about 60 % of them being usable. According to field data, more than 85 % of households were aware of the storm warning and evacuation order (vs. 60 % for Gorky in 1991), and 40 % of potentially-affected households evacuated (i.e., 3.2 million out of 8 million coastal residents).

It is difficult to generalize the case of Bangladesh on a global scale. Challenges to early warning and evacuation vary depending on the country (e.g., on the existence of transport infrastructure) and on the hazards (e.g., a storm is easier to forecast than the heavy precipitations responsible for deadly floods in Mumbai in 2005). Here, we will assume that generalizing the early warning and evacuation systems available in developed countries would make the death probability decrease from 7.5 per million to 4 per million (approximately a 50 % reduction), making it still twice as large as in developed countries.

This means that the number of lives saved every year (with current population) would be 23,000 per year. Viscusi and Aldy (2003) summarize existing estimates for the value of a statistical life (VoL) in developing countries (see Box 6.3), with values ranging from $750,000 in South Korea to $4.1 million in India.

Assuming here (in a very conservative way) that each avoided death is valued at $100,000, we obtain an annual benefit of 2.3 billion USD per year (increasing with population and economic growth). This estimate is conservative, as we do not account for morbidity (injuries and disaster-caused illness), which might play a huge role (according to EM-DAT, more than 1 million persons have been injured by weather-related natural disasters between 1988 and 2007, most of them being in Asia), and we assume that growth in population and income does not translate into more disaster-related death (which is optimistic, see for instance Kellenberg and Mobarak 2008).

**Overall, the annual benefits of improved hydromet services and early warning systems in terms of avoided human losses can thus be estimated at 2.3 billion USD per year, growing with population and economic growth.**

### *6.4.2 Economic Benefits from Hydromet Information*

Improving hydromet services would not only allow for better early warning systems. In practice, it would also produce economic benefits, in the form of useful services for industries and businesses and for households and individuals, even in normal conditions (i.e. during non-dangerous times).

In the agriculture sector, weather forecasts are used for planning purposes, e.g. to decide the dates of planting or fertilizer application. A few studies assessed the productivity gains from short to medium term weather forecasts. For instance, Wilks and Wolfe (1998) investigate the use of forecasts to optimize lettuce production in the state of New York, and they find a $900 to $1,000 gain per hectare and per year, i.e. a 10 % increase in productivity.

In the energy sector, weather forecasts are used to anticipate electricity demand, allowing us to maximize the use of lower-cost but slowly-adjusting production units (e.g., nuclear, coal, or solid biomass) and reduce as much as possible the use of higher cost production units. But weather forecasts are also used to manage production. For instance, Roulston et al. (2003) estimate the value of weather information to optimize wind power production; they find a doubling in profits thanks to 1 and 2-day forecasts.

In transport, weather forecasts are used to optimize air traffic and ship routes, and to plan road salting and other preventive actions. A study by Leigh (1995) estimates benefits from weather information at the Sydney airport. He finds a benefit of 6.9 millions of Australian dollars per year, i.e. about five million American dollars. In the United Kingdom improved meteorological information services resulted in a 20–25 % decrease in the use of road salt (Leviäkangas et al. 2007). In Croatia, the socio-economic benefit of hydromet information related to the reduced number of accidents was estimated to be 4.3–8.7 million € per year (Leviäkangas et al. 2007).

In the construction sector, weather forecasts are used to optimize the use of labor resources, and to plan some temperature- or wind-sensitive operations. It avoids accidents (e.g., a crane accident due to high wind) and improves building quality (e.g., by avoiding pouring concrete at very low temperatures).

In the tourism and health sector, weather is a predictor of future activities, useful to plan labor resources and inputs (e.g. anticipating the number of visitors in a touristic site, or the number of customers at a restaurant).

In these few examples, a significant impact of weather information on productivity is observed.[18] And the use of forecasts is growing worldwide, with many new businesses specialized in helping businesses to take this information into account. To estimate the orders of magnitude at stake, one can start from the economic value added in the world each year in the sectors considered as sensitive to weather conditions: agriculture, 2,000 billion USD; mining and energy, 7,000 billion USD; construction, 3,200 billion USD; and transport, 4,300 billion USD. Summing these numbers suggests that these sensitive sectors create more than 16,000 billion USD per year of added value, i.e. about 25 % of world GDP.

And hydro-meteorological information goes beyond weather forecasts: observations and long data series are useful to design infrastructure, building, or even urban plans. In many countries, new constructions are prohibited in flood-prone areas (e.g., the 100-year flood plain in France), but the identification of such zones can be made impossible by the absence of appropriate data (see an illustration on Casablanca and Alexandria in World Bank 2011). Hydro-meteorological information can also be used to monitor the environment over the long term, detect potentially harmful changes, and anticipate response measures (e.g., detecting as early as possible a change in rainfall to adjust drainage infrastructure, anticipating the arrival of a new pest in the agriculture sector).

Assuming that weather forecasts lead to value added gains between 0.1 and 1 % in weather-sensitive sectors, these gains would be approximately equal to between 0.025 and 0.0025 % of GDP. These estimates are small compared with the sectoral case studies presented above. And they are also small compared with estimated from ISDR et al. (2008), which propose estimates for seven countries, namely Albania, Bosnia-Herzegovina, Macedonia, Moldova, Montenegro, and Serbia and Montenegro. In these countries, an economic analysis (carried out by VTT, based on available data and detailed surveys in each of the countries) estimated benefits from hydromet services between 0.09 % (in Croatia) and 0.35 % (in Moldova).

They are also conservative because they do not account for the "small" services that hydromet information provides to each of us in our daily life. Measuring the value of the private use of forecasts is difficult. It means measuring the willingness of users to pay for the service they get from meteorological information. Knowing if one can go for a picnic without risking heavy rain has a value; being able to decide a few hours in advance if a dinner can be organized outside or inside has a value; deciding whether to take an umbrella when leaving for work in the morning also has a value. Each of these values remains small, but these decisions happen all the time and millions of people are making them. The aggregated value may thus be significant. Lazo et al. (2009) conducted a survey of U.S. households to estimate their willingness to pay for weather information that is currently provided to them, and for potential improvement of this information. The survey focused on normal

[18]It is important to acknowledge the risk of bias in the literature, published studies being the ones that find a significant impact of weather information on economic activity. To my knowledge, no paper has been published on *a lack* of such impact.

conditions, and excluded extreme events and safety aspect from the analysis, so there is no double counting with the previous section. The survey arrived at a median estimate of US$280 per year and per household, with more than 80 % of households ready to pay more than US$30. Assuming that each European household is ready to pay at least 20 euros per year, again a conservative estimate, the societal benefit from weather information would be around four billion euros per year. With a value of 80 euros, the estimate reaches 15 billion euros.

For Europe, it is thus a very conservative estimate to assume that the value of hydromet information lies between 3.4 and 34 billion euros per year. At global scale, a similarly conservative estimate is that the potential benefits would be between 16 billion and 160 billion USD per year.

As with risk mitigation, part of these benefits has already been realized. Here, we want to assess the additional benefits if the services provided in developed countries were generalized. To do so, we use the same assumption as that for risk mitigation, assuming that 10 % of these benefits are realized in low income countries, 20 % in lower middle income, 50 % in upper middle income, and 100 % in high income countries (see Table 6.2).

The result is that a generalization of hydromet services at high income country levels would generate additional economic benefits ranging between 3 and 30 billion USD. This estimate is an underestimation because it does not account for the bigger vulnerability of poor-country economies to environmental stress and weather extreme events.

It is also an underestimation because it does not account for health impact, and for the fact that higher productivity in agriculture would increase food security, with health and economic benefits.

**In that case, therefore, there are probably economic benefits that are even larger than the direct impact of early warning: economic benefits in normal conditions from hydromet information developed for disaster mitigation would yield 3–30 billion USD per year.**

### *6.4.3 How to Improve Early Warning, and at What Cost?*

Improving hydromet services and early warning systems implies multiple components. Some of the required measures are investments; some are linked to operational expenses and require sustainable funding. Other important components are linked to the institutional setting and to the training of the producers and of the users of hydromet information.

One important aspect of hydromet production information is that all countries do not need to invest in all components of the forecast chain. Remote sensing information is provided by satellites that are launched and financed by developed countries (e.g., Meteosat 5 on Africa, operated by EUMETSAT; the Metop system, also operated by EUMETSAT; NOAA 19, financed by the US government). Global modeling is also carried out regularly by many services around the world (e.g., most

**Table 6.2** Potential economic benefits from improved hydrometeorological services, and share of these benefits actually realized with current services (these benefits exclude the benefits from early warning, presented in Table 6.1)

| | GDP (million USD) | Potential (European-like) benefits | | Ratio of current vs potential benefits | Estimation of actual benefits | | Benefits from improved services | |
|---|---|---|---|---|---|---|---|---|
| | | Low estimate | Likely estimate | | Low estimate | Likely estimate | Low estimate | Likely estimate |
| Low income | 413,000 | 103 | 1,033 | 10 % | 10 | 103 | 93 | 929 |
| Lower middle income | 4,300,000 | 1,075 | 10,750 | 20 % | 215 | 2,150 | 860 | 8,600 |
| Upper middle income | 15,300,000 | 3,825 | 38,250 | 50 % | 1,913 | 19,125 | 1,913 | 19,125 |
| High income | 43,000,000 | 10,750 | 107,500 | 100 % | 10,750 | 107,500 | – | – |
| TOTAL | 63,013,000 | 15,753 | 157,533 | | 12,888 | 128,878 | **2,865** | **28,654** |

European services such as Meteo-France and MetOffice, the European center ECMWF, US institutions like NOAA). And even observation can be rationalized thanks to regional integration: a radar network is much less expensive if designed at regional rather than national scale.

What is really needed in developing countries is thus:

(i) The local observation system, based on ground, in-situ observations, for weather data (e.g., temperature, precipitation), hydrological data (runoff data), topological data (e.g., elevation database to link runoff forecast with flood extension), socio-economic data (e.g., population density and transportation capacity to decide on evacuation).
(ii) Forecasting capacity, i.e. the translation of low-resolution model forecast into high-resolution forecast, using statistical downscaling and correction or additional models.
(iii) Interpretation capacity, to translate model output into actual forecast and warnings, taking into account known model bias and specific local conditions that models cannot integrate.
(iv) Communication tools, to make sure the alert reaches the individuals in charge of implementing prevention measures (including the public who is supposed to evacuate), and evacuation and emergency plans.
(v) Users' decision-making capacities, to make sure warnings are actually used (including for evacuation).

The ability to improve forecast capacity, and to provide forecasts that are appropriate and used on the ground, will depend on the local scientific "capacity", including the existence of university and research programs. An example of how capacity can be improved is provided by the AMMA program (African Monsoon Multi-disciplinary Analysis[19]), an international research program that resulted in (i) better observation systems in Africa; (ii) creation of local research teams on hydrometeorology, health and agriculture; (iii) creation of new degrees in relevant disciplines in African universities.

Finally, the ability to implement prevention measures will depend on local infrastructure and capacity. Lack of transport capacity (when public transportation is not accessible and poor households do not have individual transport vehicle) can make the entire system less efficient. Also, limited investments can make it possible to implement prevention measures, but financial constraints may make them impossible. For instance, hard-protection to windows and cofferdam can be extremely efficient to reduce wind or flood damages to houses and cost only a few dozen USD. They may nevertheless be too expensive to be broadly available in some regions.

The five categories of action may also be designed very differently depending on the local contexts. For instance, communication through TV, radio and cell phones can be extremely efficient, but in some countries low-tech options (such as the bicycle mounted loudspeakers used in Bangladesh) can be useful complements.

[19]More information on http://www.amma-international.org.

When trust does not exist between the population and forecast producers, local contact points (such a Red Cross volunteers) can help in building it.

Here, we focus on the cost of the five elements listed above, without consideration for the more general capacity in the country, assuming that it will not be a constraint on evacuation efficiency (keeping in mind that we still assume that even with early warning and evacuation systems, the death ratio will be twice as large in developing countries as in developed countries, because of lower building quality and transportation infrastructures).

The cost of strengthening hydromet service depends on the size of the country and of the context.[20] The range can be from a few million USD to hundreds of millions. The NMS can be as small as 5–8 forecasters converting results of global forecasting centers in the national alerts and forecasts supported by very limited infrastructure in the country. In this case the operating costs can be comparable or exceed the investment costs of setting up the system. This is not a typical case. Usually, in developing countries there are many dozens (or several hundred) of low paid staff dealing with extensive but dilapidated infrastructure (buildings, observation sites, instruments, etc.).

In Russia, the first hydromet project supported by the World Bank cost over $172 M. The second (currently under preparation) is estimated at $141 M. Still this is a fraction of the total needs (and real costs). In Mexico, the World Bank is planning to invest $109 M. Recent investments in Poland and Turkey were about $62 M and $26 M. According to an analysis on South Eastern Europe (ISDR et al. 2008), the estimated financing needed to strengthen national hydromet services in seven countries, without regional cooperation and coordination, would be around €90.3 million. With deeper cooperation, the cost is around €63.2 million. This cost includes investment for in-situ measurement, upper-air sounding, radars, communication, and dissemination of data, information, forecast and warning. It includes also maintenance over a 5-year period.

Moreover, as demonstrated by the case of Bangladesh, forecast capacity is not sufficient and needs to be complemented by a decision-making system to translate a forecast into a warning, and by communication tools (possibly with volunteers on the ground to diffuse the evacuation orders).

To be conservative, we will thus assume that the cost of providing appropriate early warning and evacuation orders in developing counties would be equal to about $50 million per country over a 5-year period (including maintenance and operational costs). For a set of 80 developing countries, including most of the world population, the cost would be equal to $2 billion over 5 years, i.e. an annual cost of $800 million per year.

It is important for hydro-meteorological services to be well connected to their users to ensure they produce the information that is required for early warning and evacuation (and for operational management). Also, users of information need to

[20]Invaluable information has been provided by Vladimir Vtsirkunov, from World Bank/GFDRR.

know how to use the information, especially for forecasts that are always uncertain. Deciding when to evacuate a flood-prone area necessitates an analysis of the cost of a false alarm (including the loss of confidence, which can lead the population not to evacuate at the next alert), the cost of missing a dangerous flood, and the probability of error. The easiest way to achieve this is to include hydro-meteorological services in government services that will use hydro-meteorological information and decides on evacuation and preparation measures (e.g., the Ministry of Interior in most countries).

Producing hydromet information and making decisions on when to evacuate and invest in preparation measures requires specific skills that are not always available in developing countries. In addition to investments in hydromet and risk management services themselves, it is thus necessary to invest in population skills, for instance through the creation of a specific university program that will train students on hydromet information production, risk management, and decision-making. Training about 40 individuals per year in 40 countries would cost less than $200 million per year.

### *6.4.4 Conclusions on Investments in Hydrometeorological Information and Early Warning*

This analysis demonstrates the large potential of investments in hydro-meteorological services and early warning and evacuation schemes to reduce the human and economic losses due to natural disasters. It also stresses the existence of other significant socio-economic benefits, which can potentially exceed the benefits in terms of disaster risk reduction.

**A ballpark estimate of the cost of improvement of hydromet and warning services to make it possible to implement early warning systems in developing countries is thus lower than $1 billion a year. With benefits in terms of disaster risk reduction between $2.6 and $4.3 billion per year, and economic benefits between $3 and $30 billion, we reach benefit-cost ratios between 5 and 35.**

Since this analysis compares annual investment and spending to annual benefits, it does not depend on the discount rate. There might however be a several-year delay between an increase in spending and the corresponding increase in benefits. But such a delay should not change in a qualitative manner the result of the analysis proposed here (Table 6.3).

This analysis is done at global scale, using very simple assumptions that are only capable of providing orders of magnitude. Before real investments are made, local and context-specific analyses are necessary, at project scale. In some regions, for instance, the main risk is linked to heavy precipitation, and meteorological radars will be extremely useful to provide warning with a time-lead of a few hours. Such radars will be much less needed where risks are not related to heavy precipitation but – say – to droughts or temperature extremes.

**Table 6.3** Summary of benefits from and costs of upgraded hydro-meteorological services

| Type of benefits | Annual benefits (million USD) | | Annual cost (million USD) | Benefit-cost ratio | |
|---|---|---|---|---|---|
| | Minimum | Maximum | | Minimum | Maximum |
| Reduced asset losses from disasters | 300 | 2,000 | 1,000 | 5.6 | 34 |
| Reduced human losses from disasters | 2,300 | | | | |
| Other economic benefits | 3,000 | 30,000 | | | |
| Total | 5,600 | 34,300 | | | |

**Even though the implementation would need to be done country by country – with detailed analyses of each investment – one can make a strong case for increased investment (and international attention) in early warning systems to save lives and improve economic efficiency.**

## 6.5 Conclusion

Building a flood protection system able to protect a city against large scale disaster (such as a Category 5 hurricane in New Orleans) or a hydro-meteorological system to observe and predict weather extreme events is a huge investment, and it is wise to assess costs and benefits before any implementation decision: other projects might be more efficient to save lives and improve the population's well-being. In some cases, like in the case of early warning systems, a CBA using conservative assumptions is able to demonstrate that an investment is largely cost-effective.

One must, however, be careful with the results. The case study on New Orleans shows that using probabilities derived from historical experience and direct cost estimates produced by insurance companies lead to an underestimation of benefits. **Generally, a CBA is useful but should encompass the whole set of possible assumptions to check its robustness. A CBA can reach very different results for reasonable parameter values and therefore can rarely be used to make a decision in an objective way.**

**It is crucial to note that result uncertainty does not arise only from scientific uncertainty (e.g., on climate sensitivity or building vulnerability) that could be reduced with more research. Most of it arises from value judgments on which consensus exists or is likely to exist in the future. Since uncertainty is very large and ethics considerations are important, risk management decisions will remain political ones.**

**CBA, however, can be used to collect information on the consequence of a project, and to help organize the debate, by linking the different opinions of various groups on what should be done to different opinions about the parameters of the analysis (e.g., the discount rate, or the amount of avoidable losses). CBA should therefore be understood as a complement and a tool to open consultations and discussions, not as a replacement for them.**

# References

Ackerman F, DeCanio SJ, Howarth RB, Sheeran K (2009) Limitations of integrated assessment models of climate change. Clim Chang 95(3–4):297–315

ADB (2007) Theory and practice in the choice of social discount rates for cost benefit analysis: a survey. Asian Development Bank, Manila

Allenby B, Fink J (2005) Toward inherently secure and resilient societies. Science 309:1034–1036

Arrow KJ (1950) A difficulty in the concept of social welfare. J Polit Econ 58(4):328–346

Arrow KJ (1951) Social choice and individual values. Wiley, New York

Arrow KJ (1974) Chapter 3: The theory of risk aversion. In: Essays in the theory of risk-bearing. North-Holland, Amsterdam

Arrow KJ, Lind RC (1970) Uncertainty and the evaluation of public investment decisions. Am Econ Rev 60:364–378

Arrow KJ, Cropper ML, Eads GC, Hahn RW, Lave LB, Noll RG, Portney PR, Russel M, Schmalensee RL, Smith VK, Stavins RN (1996) Benefit-cost analysis in environmental, health, and safety regulation. American Enterprise Institute Books and Monographs, Washington, DC

Arrow KJ, Cropper ML, Gollier C, Groom B, Heal GM, Newell RG, Nordhaus WD, Pindyck RS, Pizer WA, Portney PR, Sterner T, Tol RSJ, Weitzman ML (2013). How should benefits and costs be discounted in an intergenerational context? Resources for the future discussion paper 12, p 53. Washington, DC, USA

Barredo JI (2009) Normalised flood losses in Europe: 1970–2006. Nat Hazards Earth Syst Sci 9:97–104

Bentham J (1789) An introduction to the principles of morals and legislation. Payne, London; republished, Clarendon Press, Oxford, 1907

Burby RJ, Nelson AC, Sanchez TW (2006) The problem of containment and the promise of planning. In: Birch EL, Wachter SM (eds) Rebuilding urban places after disaster, lessons from Hurricane Katrina. University of Pennsylvania Press, Philadelphia

Carsell KM, Pingel ND, Ford DT (2004) Quantifying the benefit of a flood warning system. Nat Hazards Rev 5:131–140

Day HJ (1970) Flood warning benefit evaluation-Susquehanna River basin (urban residences), ESSA technical memorandum WBTM Hydro-10. National Weather Service, Silver Spring

Dercon S (2007) Fate and fear: risk and its consequences in Africa. Paper prepared for the African Economic Research Consortium. Oxford University, Oxford

Fafchamps M, Udry C, Czukas K (1998) Drought and saving in West Africa: are livestock a buffer stock? J Dev Econ 55(2):273–305

Gollier C (2002) Time horizon and the discount rate. J Public Econ 85:463–473

Gollier C (2008) Discounting with fat-tailed economic growth. J Risk Uncertain 37:171–186

Gollier C (2012) Pricing the planet's future: the economics of discounting in an uncertain world. Princeton University Press, Princeton

Golnaraghi M (2012) National hydro and meteorological services. Presentation at the 9th international forum of the meteorology. Geneva, 22 Mar 2012

Hallegatte S (2006) A cost-benefit analysis of the New Orleans flood protection system. AEI-Brookings Joint Center. Regulatory analysis 06-02. March 2006. Washington, DC, USA

Hallegatte S (2008) A proposal for a new prescriptive discounting scheme: the intergenerational discount rate. FEEM working paper No. 47.2008. Milan, Italy

Hallegatte S (2012) A cost effective solution to reduce disaster losses in developing countries: hydro-meteorological services, early warning, and evacuation. Policy Research working paper 6058, The World Bank, Washington, DC, USA

Hanson S, Nicholls R, Ranger N, Hallegatte S, Corfee-Morlot J, Herweijer C, Chateau J (2011) A global ranking of port cities with high exposure to climate extremes. Clim Chang 104(1):89–111

Harberger AC (1978) On the use of distributional weights in social cost–benefit analysis. J Polit Econ 86(2):S87–S120

Harberger AC (1984) Basic needs versus distributional weights in social cost-benefit analysis. Econ Dev Cult Chang 32(3):455–474

Heal G (2009) Climate economics: a meta-review and some suggestions for future research. Rev Environ Econ Policy 3(1):4–21

International Commission for the Protection of the Rhine (2002) Non structural flood plain management – measures and their effectiveness. ICPR, Koblenz

IPCC (2013) Climate change 2013: the physical science basis. Contribution of Working Group I to the fifth assessment report of the Intergovernmental Panel on Climate Change [Stocker TF, Qin D, Plattner G-K, Tignor M, Allen SK, Boschung J, Nauels A, Xia Y, Bex V, Midgley PM (eds)]. Cambridge University Press, Cambridge/New York, 1535 pp

ISDR, World Bank, WMO, IMF (2008) Strengthening the hydro-meteorological services in South Eastern Europe. Available on http://www.unisdr.org/we/inform/publications/7650

Kazianga H, Udry C (2006) Consumption smoothing? Livestock, insurance, and drought in rural Burkina Faso. J Dev Econ 79(2):413–446

Kellenberg DK, Mobarak AM (2008) Does rising income increase or decrease damage risk from natural disasters? J Urban Econ 63:788–802

Khan MR, Rahman A (2007) Partnership approach to disaster management in Bangladesh: a critical policy assessment. Nat Hazards 41(1):359–378. doi:10.1007/s11069-006-9040-y

Kreibich H, Muller M, Thieken AH, Merz B (2007) Flood precaution of companies and their ability to cope with the flood in August 2002 in Saxony, Germany. Water Resour Res 43(3):1–15

Kreibich H, Thieken AH, Petrow TH, Muller M, Merz B (2005) Flood loss reduction of private households due to building precautionary measures – lessons learned from the Elbe flood in August 2002. Nat Hazards Earth Syst Sci 5:117–126

Lazo JK, Morss RE, Demuth JL (2009) 300 billion served: sources, perceptions, uses, and values of weather forecasts. Bull Am Meteorol Soc 90(6):785–798

Leigh RJ (1995) Economic benefits of Terminal Aerodrome Forecasts (TAFs) for Sydney airport, Australia. Meteorol Appl 2:239–247

Leviäkangas P, Hautala R, Räsänen J, Oörni R, Sonninen S, Hekkanen M, Ohlström M, Venäläinen A, Saku S (2007) Benefits of meteorological services in Croatia. VTT TIEDOTTEITA 2420

Mankiw G (1981) The permanent income hypothesis and the real interest rate. Econ Lett 7:307–311

Marshall A (1890) Principles of economics, 9th edn. Macmillan, London, 1961

Mechler R (2005) Cost-benefit analysis of natural disaster risk management in developing countries. Deutsche Gesellschaft für Technische Zusammenarbeit (GTZ), Eschborn

Merz B, Kreibich H, Thieken A, Schmidtke R (2004) Estimation uncertainty of direct monetary flood damage to buildings. Nat Hazards Earth Syst Sci 4:153–163

Muir-Wood R, Grossi P (2006) Flood risk in New Orleans: implications for future management and insurability, risk management solutions. Available on www.rms.com

Paul BK (2009) Why relatively fewer people died? The case of Bangladesh's Cyclone Sidr. Nat Hazards 50:289–304

Pigou AC (1920) The economics of welfare. Macmillan, London

Ramsey FP (1928) A mathematical theory of saving. Econ J 38(152):543–559

Rawls J (1971) A theory of justice. Harvard University Press, Cambridge

RMS (2005) Hurricane Katrina: profile of a super cat. Lessons and implications for catastrophe risk management. https://www.rms.com

Robbins L (1938) Interpersonal comparisons of utility: a comment. Econ J 48(192):635–641

Roulston MS, Kaplan DT, Hardenberg J, Smith LA (2003) Using medium-range weather forecasts to improve the value of wind energy production. Renew Energy 28:585–602

Skoufias E, Quisumbing A (2003) Consumption insurance and vulnerability to poverty: a synthesis of the evidence from Bangladesh, Ethiopia, Mali, Mexico and Russia. Manuscript, IADB and IFPRI

Stern N (2006) The economics of climate change: the stern review. Cambridge University Press, Cambridge

Subbiah A, Bildan L, Narasimhan R (2008) Background paper on assessment of the economics of early warning systems for disaster risk reduction. World Bank Group for disaster reduction and recovery. http://gfdrr.org/gfdrr/sites/gfdrr.org/files/New%20Folder/Subbiah_EWS.pdf

Swiss Re (2006) The effect of climate change: storm damage in Europe on the rise. http://www.preventionweb.net/files/20629_publ06klimaveraenderungen1.pdf

Thieken AH, Kreibich H, Muller M, Merz B (2007) Coping with floods: preparedness, response and recovery of flood-affected residents in Germany in 2002. Hydrol Sci 52(5):1016–1037

U.K. Treasury (2003) Green book, appraisal and evaluation in Central Government. Available on http://greenbook.treasury.gov.uk/

Viscusi WK, Aldy J (2003) The value of a statistical life: a critical review of market estimates throughout the world. J Risk Uncertain 27(1):5–76

Weitzman ML (1998) Why the far-distant future should be discounted at its lowest possible rate. J Environ Econ Manag 36:201–208

Weitzman M (2001) Gamma discounting. Am Econ Rev 91(1):260–271

Wiener JB (1998) Managing the iatrogenic risks of risk management. Risk Health Saf Environ 9(39):40–82

Wilks DS, Wolfe DW (1998) Optimal use and economic value of weather forecasts for lettuce irrigation in a humid climate. Agric For Meteorol 89:115–130

Williams BA (2002) Fran, Floyd and mitigation policy. Berry A. Williams and Associates, Inc. Collinsville, IL, USA

World Bank (2011) Climate change adaptation and natural disasters preparedness in the coastal cities of North Africa. The World Bank, Washington, DC, USA

# Chapter 7
# Decision Making for Disaster Risk Management in a Changing Climate

**Abstract** The uncertainty in future climate conditions means that new decision-making approaches need to be mobilized. This chapter highlights strategies that are best able to cope with the large uncertainty (and disagreement) that surround the climate change impacts on natural hazards and disasters. It introduces the robust decision-making approach and suggests flexible and reversible strategies that are best suited to the current situation of high uncertainty on future climates. These approaches can be used to improve risk management, and capture the potential synergies between risk management and climate change adaptation.

**Keywords** Decision-making • Uncertainty • Robust decision-making • Disaster risk management

The limits of the CBA analysis proposed on the New Orleans case in the previous chapter will also be met for many other decisions that need to be made in the following decades. Indeed, many decisions have a very long time horizon, and are sensitive to climate conditions.

For instance, most buildings are supposed to last up to 100 years and will have to cope in 2100 with climate conditions that, according to most climate models, will be radically different from current ones. So, when designing a building, architects and engineers have to be aware of and account for the future changes that can be expected. Taking the example of Paris, many climate models project that the Paris climate in 2070 – in the absence of ambitious mitigation policies – will be comparable to the climate that can be found today in the South of Spain. Obviously, buildings are designed differently in Paris and in the South of Spain, and this should be taken into account. Building norms that are applied today in France to reduce heating energy consumption in winter may even become counter-productive if they do not take into account the risk of high energy consumption from air conditioning in summer in a few decades from now. The adaptation of water policy leads to the same issue: Milly et al. (2008) demonstrate why water management cannot keep using the stationary hypothesis in its investment decisions, as water

---

Part of this chapter is based on Hallegatte (2009), published in Global Environmental Change, and Hallegatte et al. (2012).

 
S. Hallegatte, *Natural Disasters and Climate Change*,
DOI 10.1007/978-3-319-08933-1_7

**Table 7.1** Sectors for which climate change should be taken into account as of now, because of time scale or sensitivity to climate conditions. Sensitivity is estimated by the author

| Sector | Time scale (year) | Sensitivity |
|---|---|---|
| Water infrastructures (e.g., dams, reservoirs) | 30–200 | +++ |
| Land-use planning (e.g., in flood plain or coastal areas) | >100 | +++ |
| Coastline and flood defenses (e.g., dikes, sea walls) | >50 | +++ |
| Building and housing (e.g., insulation, windows) | 30–150 | ++ |
| Transportation infrastructure (e.g., port, bridges) | 30–200 | + |
| Urbanism (e.g., urban density, parks) | >100 | + |
| Energy production (e.g., nuclear plants) | 20–70 | + |

Source: Hallegatte (2009)

trends (rainfall, runoff, etc.) are changing in response to climate change. Since they report that more than US$500 billion are invested every year in this sector, the implementation of new practices cannot be delayed: water management needs to use climate model outputs to optimize strategies and policies. Table 7.1 identifies a few sectors where this problem is particularly acute.

Fortunately, there has been a significant rise in worldwide awareness about climate change. The positive outcome of this shift in awareness is that many architects, urban planners, water managers, and other planners are now concerned about how climate change will influence their activities, and a spontaneous adaptation is likely to occur.

**This new awareness is not enough. Climate change represents more than just a change in climate conditions. For decision-makers, climate change represents, more importantly, a dramatic increase in uncertainty.** In the past, the climate parameters pertinent to most activities could be observed and measured. In presence of well-posed objectives, statistical analyses and optimization algorithms were able to produce "best" designs as a function of known climate conditions (e.g., dike heights as a function of the return time of certain storm surges, or building characteristics as a function of typical temperature levels). In the future, however, substantial climate uncertainty makes such methods more difficult to apply. As an illustration of this uncertainty, it is enough to recall that climate models still disagree *on the sign* of precipitation changes in West Africa. It means that while water management needs to take into account future changes in climate conditions, we are unable to say whether this change will be an increase or a decrease in rainfall. Clearly, this uncertainty represents an important obstacle to the design of water management.

**To cope with this uncertainty, new decision-making methods have to be developed and used. Some of these methodologies are based on sequential decision-making and cost-benefit analysis under uncertainty.** But these methods require a quantified assessment of the likelihoods of the different possible outcomes, i.e. subjective probabilities. In most cases, these quantified likelihoods cannot be estimated (see previous chapter, Dessai et al. 2009; Hall 2007). Even when some methodologies can be used to assess subjective probabilities, they often fail to create a consensus. A cost-benefit analysis can only be performed if the various parties to a decision agree on the many "parameters" of the analysis: the discount rate, the probability of various possible futures and events (e.g., the probability that climate change leads to an increase in hurricane intensities), the value of different non-markets impacts (e.g., the value of the loss of a forest or a species), etc. But these parameters are highly dependent on political choices, world views, values, and beliefs, and one cannot expect a consensus on their values. These parameters involve what can be referred to as "deep uncertainty".

**Deep uncertainty refers to situations in which experts cannot agree on which models to use; on the probability distributions of key uncertain parameters; or on the values of alternative outcomes.** In such cases, it is not possible to define a subjective probability for alternative outcomes, or even to identify the set of possible futures (including highly improbable events – like the famous "black swan"). And different views cannot be reconciled through a common estimation of probabilities of different outcomes. A situation of deep uncertainty is different from a situation of "large uncertainty," in which different actors can agree on the probabilities and values of different outcomes, even if the range of possible outcomes is very broad because knowledge is limited. In situations of deep uncertainty, different stakeholders or experts can have divergent opinions and may not even agree on the existence of large uncertainty. Deep uncertainty often leads to strong political opposition to any action, and therefore to policy paralysis.

**Because the cost-benefit analysis requires all parties to agree on these parameters before the analysis starts, it often leads to deadlocks and paralysis in situations of deep uncertainty. This is the case for all methods that we can label "agree-and-analyze", in which the parties have to agree on the basic assumptions of an analysis before it is performed.** In these cases, it is likely that the discussion among parties will stall at the parameter or definition stage, and will be unable to provide useful insight to bring the parties together and make them reach a consensus.

**Other methods are based on an "analyze-and-agree" sequence, in which the analysis allows for multiple worldviews, beliefs and values to be included. These methodologies try and bring the need to agree at the end of the analysis, and to restrict the need for a consensus to the decisions to be made, and not to the parameters that drive the decision**. In other terms, they try to help the parties to a decision to agree on what to do (e.g., whether to build a dike or not) regardless of their beliefs on climate change, instead of making them agree on climate change itself.

**These approaches also put the emphasis on the need to avoid irreversible choices that can lead to lock-ins in undesirable or unacceptable situations.** For instance, developing low-lying coastal areas with the current uncertainty on future sea level rise is dangerous, as it can lead to a difficult choice between expensive protection and retreat in the future. If such a situation is considered undesirable or unacceptable, and while the uncertainty is so large, it is preferable to develop only the areas that are not too vulnerable in case pessimistic sea level rise scenarios are confirmed. This is the basic concept of robust decision-making.

## 7.1 Methodologies for Robust Decision-Making

**To avoid irreversible choices that can lead to large regrets, it is possible to base decisions on scenario analysis and to choose the most robust solution, i.e. the one that is the most insensitive to future climate conditions, instead of looking for the "best" choice under one scenario** (e.g., Schwartz 1996; Lempert et al. 2006; Lempert and Collins 2007). For professionals, these methods are consistent with those commonly used to manage exchange-rate risks, energy cost uncertainty, research and development outcomes, and many other situations that cannot be forecast with certainty. Such robust decision-making methods have already been applied in many long-term planning contexts, including water management in California (Groves and Lempert 2007; Groves et al. 2007). For most decision-makers, the novelty will be the application of these methods to climate conditions. This requires users of climate information to collaborate more closely with climate scientists and to adapt their decision-making methods to the climate change context.

### *7.1.1 Robust Decision-Making*

**Robust decision making** (Lempert and Collins 2007) **rests on three key concepts that differentiate it from the traditional subjective expected utility decision framework:**

- **multiple views of the future,**
- **robustness criterion, and**
- **iterative process based on a vulnerability-and-response-option rather than a predict-then-act decision framework, and focus on strategies that evolve over time in response to new information.**

**First, like traditional scenario methods, RDM characterizes uncertainty with multiple views of the future.** These multiple views are represented analytically by multiple future states of the world. RDM can also incorporate probabilistic information, but rejects the view that a single joint probability distribution represents the best description of a deeply uncertain future. Rather RDM uses ranges, or more formally sets, of plausible probability distributions to describe deep uncertainty. As described below, considering multiple views of the future can help include all stakeholders in a decision, promote consensus, and reduce the tendency to underestimate uncertainty.

**Second, RDM uses a robustness rather than an optimality criterion to assess alternative policies.** The traditional subjective utility framework ranks alternative decision options contingent on the best estimate probability distributions. In general there is some best (i.e. highest ranking) option. The shortcoming of the optimal solution is that it is only optimal for the predicted future and may be poor otherwise. A robustness criteria, in contrast, seeks solutions that are good

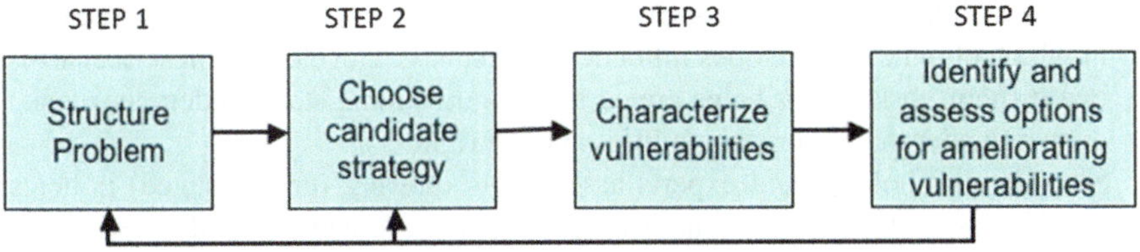


**Fig. 7.1** Steps in Robust Decision Making (RDM) analysis

(though not necessarily optimal) no matter what the future. There exist several specific definitions of robustness, but all incorporate some type of satisfying criteria. For instance, a robust strategy can be defined as one that performs reasonably well compared to the alternatives across a wide range of plausible future scenarios. Often there is no single robust strategy but a set of reasonable choices that decision makers can choose among.

One of the most effective means to achieve a robust strategy is to explicitly design it to evolve over time in response to new information. An RDM analysis often pays particular attention to simulating the evolution over time of not only the climate and other bio-physical systems, but also of the policy itself as it responds to a wide variety of future contingencies.

**Third, RDM employs a vulnerability-and-response-option analysis framework to characterize uncertainty and to help identify and evaluate robust strategies,** as summarized in Fig. 7.1. In the first step, analysts structure the problem and assemble the relevant data and simulation models similarly to most decision analyses. The next step differs significantly from a predict-then-act analysis. The latter analyses begin by characterizing uncertainty about the future either with a point forecast, probabilistic forecasts, or multiple scenarios. In contrast, RDM begins by considering one or more candidate strategies. Depending on the situation, this strategy(ies) might derive from a variety of sources. In some cases the initial candidate strategy might be current policy. In other cases, the initial candidate strategy might be one of several new policies proposed by an agency or other stakeholders in the public debate. In yet other cases, analysts might perform a traditional analysis and use the resulting optimum strategy as the initial candidate for the RDM analysis; in that case, the robust decision-making analysis is a complement to the traditional analysis, to check the robustness and vulnerability of the retained strategy.

To summarize, a RDM approach can be implemented through the following steps (see Fig. 7.1):

- **Step 1:** stakeholders have to agree on a set of possible scenarios or on a range of possible values for the various parameters. The objective is to cover as broadly as possible the range of possible futures. In that process, stakeholders do not have to agree on what is most likely or plausible; they only have to accept considering the scenarios proposed by other stakeholders (e.g., someone who thinks climate change will not impact hurricane intensities has to accept to investigate a scenario in which climate change does influence hurricanes). Developing these scenarios is a difficult task, but it helps create a dialogue among stakeholders and make them consider that other stakeholders may be right.

- **Step 2:** stakeholders and experts assess how existing (or envisaged) policies would cope with all these scenarios, possibly using numerical models. Doing so requires running models or quantitative analyses to investigate the consequences of the analyzed policy in each of the scenarios, which can represent a large amount of work (and imply heavy computer-based simulations). In an ideal context, the analyses should rely on multiple models to take into account not only the uncertainty on parameters, but also on modeling choices.
- **Step 3:** Stakeholders need to decide which scenarios are considered as "failures", i.e. as cases where the policy or the plan does not provide a satisfying outcome. Again, the analysis does not require stakeholders to agree on a unique metrics to assess the outcome of a policy in a given scenario, but only on what should be considered as success or failure.

  This characterization can use one or several metrics, including classical economic metrics such as the cost-benefit ratio.[1] But in addition, distributional impacts or health consequences can be easily accounted for. In some situations, the strategy's robustness might be assessed using an absolute performance criterion. For instance, the strategy fails in a particular case if the cost or impacts on human health exceed some threshold amounts. In other situations, robustness might be assessed using a relative performance criterion, such as deviation from optimality (also called "regret"). For instance, for each entry in the database, analysts might compare the cost of the candidate strategy to the least cost strategy that achieves some human health target. Policy makers might consider the candidate strategy a failure in any given case if its cost exceeds the least cost by some amount. Both the absolute or relative performance measures represent satisfying criteria.
- **Step 4:** the RDM analysis identifies alternative strategies that might ameliorate the vulnerabilities of the candidate strategy and summarizes the tradeoffs such that policy makers can decide whether or not to adopt one of these alternative strategies. If one or several scenarios fail, the question is whether it would be possible to change course over time when it appears clear that these scenarios will materialize, in order to avoid such unacceptable outcomes. Policy makers could then use this information to decide whether or not to adopt the new strategy. If they do so choose, analysts repeat the process starting at the second step using this new candidate strategy. Alternatively, policy makers might be dissatisfied with both their original and new options, and using what they have learned about vulnerabilities and potential responses return to the first step to reformulate their decision problem in a way that might yield more desirable robust options.

[1]Because a robust decision-making process can take the result of a cost-benefit analysis as an input, it should not be considered as an alternative to the cost-benefit analysis, but as a complement.

Of course, it can happen that no solution is robust in all possible futures. In that case, stakeholders need to determine whether the residual vulnerability represents an acceptable level of risk-taking, or not.

This method helps build policy mixes that are robust to most possible futures, provided that scenarios are well defined. It aims at avoiding lock-ins in undesirable situations in some scenarios, and at exploring the range of possible futures, including unlikely scenarios (Box 7.1).

**Box 7.1: Illustrative Example of How Robust Decision-Making Can Be Applied to Water Management (Adapted from Groves et al. 2007)**
The Inland Empire Utilities Agency (IEUA) provides water to 800,000 people in California, from several sources: groundwater, imported water, recycled water, surface water, and desalted water. To analyze the 25-year plan of the IEUA and its vulnerability to climate change, the IEUA and the Rand Corporation carried out an analysis based on the robust-decision-making framework with the following steps:

- Step 1: Two climate scenarios were considered, with a more or less adverse consequence of climate change on water availability. Additional uncertainties were included to build 200 scenarios, including uncertainties in future water use efficiency, on the cost of imported water, and on the existence of groundwater replenishment and water-recycling programs.
- Step 2: An analysis of water shortage risks in all scenarios demonstrated the risk of unacceptable costs in 120 out of the 200 scenarios. The largest sources of uncertainty are the future precipitation levels and the introduction of aggressive water recycling policies.
- Step 3: The vulnerabilities of existing plans are identified, with the scenarios in which they are likely to occur. The results suggest that existing plans would fail if there is a combination of large decrease in precipitation and low efficiency of existing water recycling programs.
- Step 4: The analysis considered several policy options to cope with these unacceptable possibilities, including some with immediate actions and some with delayed actions when more information is available. Considering the potential costs in the future, IEUA decided to accelerate immediately its water recycling program, since it reduces the likelihood of unacceptable scenarios in return for a small immediate cost.

The robust decision-making approach made the IEUA change its strategy, trading higher short-term costs against lower future vulnerability to undesirable scenarios.

### 7.1.2 Advantages over Other Approaches

**One of the advantages of this methodology is to identify the uncertainties that matter, and to help the analysis focus on them.** For instance, the analysis can determine that some large uncertainty does not matter, and that it is thus useless to spend time and effort to build a consensus or to collect more data on the topic. An analysis of flood risks in Ho Chi Minh City in Vietnam (Lempert et al. 2013) investigated the performance of more than ten flood management plans, taking into account multiple deep uncertainties, such as future population, economic growth, and the effects of climate change on rainfall and the sea level.

The analysis found that the current infrastructure plan reduces risk in best-estimate future conditions and that it is robust to a wide range of possible future population and economic trends. It means that even though there is a large uncertainty on future economic and population growth, it is not necessary to spend time and effort to try to predict and agree on future economic conditions and populations to develop a flood management plan: more information on this topic is not useful in the design of the plan (or more precisely, the plan is robust to the uncertainty we have on this information). On the other hand, the current plan was found to be vulnerable to large changes in precipitations or water sea level, and the uncertainty on these factors is critical to design a flood management plan. **There is a high value in identifying the factors that have the largest influence on the failure or success of a plan, since it helps focus on them the efforts in data collection, modeling, and stakeholder discussions. It also helps avoiding a deadlock relative to an uncertainty that eventually does not matter for the decision that needs to be made in a given context.**

**Another advantage is the possibility to include multiple metrics in the definition of what is a failure or a success.** For instance, as already stated, classical economic criteria such as the benefit-cost ratio, the internal rate of return or the net present value can be included in a robust decision-making. But these metrics can be complemented with other indicators that are difficult to monetize, such as distributional impacts (e.g., the impact on the poorest), environmental impacts, health impacts (including casualties and fatalities), or cultural losses or benefits. This possibility, and the fact that the methodology does not require ranking of all options using a single metrics, reduces the need for consensus and agreement on values and preferences, making the robust decision-making approach easier to apply when decisions are highly controversial and political.

Nevertheless, the application of robust decision-making strategies to real-world cases is not without difficulty. The construction of scenarios is a long and work-intensive process, which requires the involvement of many stakeholders. Because it forces stakeholders to discuss possible futures they may not have considered before, however, the scenario-building process is in itself an important learning tool. It can help decision-makers and stakeholders revise their own choices toward higher resilience and lower vulnerability (Box 7.2).

**Box 7.2: Robust Decision-Making in Everyday Life**

Many decisions that each of us has to make every day also depends on very uncertain parameters. And the way we make these decisions is sometimes very close to what a robust decision-making framework would suggest.

Consider for instance the way most individuals make the decision of buying home fire insurance. Few would use the probability of a home fire to make this decision. Instead, the question everybody asks himself is "what would I do if I have a fire in my home?" If the consequences are limited, for instance thanks to high savings, one may decide he does not need fire insurance. But if the consequences are dire, and if the price of fire insurance is low enough, most people decide to buy home insurance regardless of the probability of such an event.

In this example, the decision-making creates scenarios for the future (with and without fire, with and without insurance), determine which scenarios lead to unacceptable outcomes (here, a fire without insurance), and assess the cost of avoiding this vulnerability (i.e. the cost of buying insurance). Considering that the cost of this robustness is low, most individuals buy fire insurance. But the role of probability in the decision is much lower than the role of the consequences in each of the scenarios (with or without a fire).

**The Robust Decision-Making approach is thus more than a decision-making tool; it is a decision-making process that builds on stakeholders' knowledge, and helps them reach an agreement. In situations of deep disagreement between various stakeholders, it can also be considered as a negotiation support. Because it forces each stakeholder to envisage the possibility that other stakeholders are right, it helps build a constructive dialogue and reach a widely-acceptable consensus.**

In applying robust-decision-making, one crucial element is the ability of learning in the future, and to adjust policies and measures as a function of new information. This possibility is dependent upon (i) a data collection and research and development process, to collect additional information over time; (ii) an institutional scheme that allows regular updates of policies and measures, accounting for new information. An example is the compulsory 5-year revision of water policy in the Netherlands, which forces the various actors and decision-makers to take into account the most recent science and the latest socio-economic developments in Dutch water policies.

**More than a unique decision, at one point in time, robust decision-making should be understood as a learning-and-acting process that takes place over time. It thus requires first the creation of the institutions that will carry out this long-term process. As it has been regularly mentioned, strategies for climate change adaptation and disaster risk reduction should include a strong institutional component.**

## 7.2 Robust Strategies for Disaster Risk Management

Robust decision-making approaches tend to favor strategies that are more flexible and adapted to high-uncertainty situations. To help design such strategies, one can identify a few broad categories of strategies that are likely to be better able to cope with uncertainty (see also Fankhauser 1994; Fankhauser et al. 1999). This list does not pretend to be comprehensive, but to help decision-makers find innovative and flexible solutions. Table 7.2 provides examples of adaptation measures, and rank them according to their ability to cope with uncertainty.

### *7.2.1 No-Regret Strategies*

"No-regret" measures constitute a first category of strategies that are able to cope with climate uncertainty. These strategies yield benefits for all possible climate changes, and even in absence of climate change.

For example, controlling leakages in water pipes is almost always considered a very good investment from a cost-benefit analysis point-of-view, even in the absence of climate change. On the other hand, additional irrigation infrastructure is an interesting measure in some regions in the current climate. In others, considering the high investment costs necessary, it would be beneficial only if climate change changes precipitation. So, irrigation is a no-regret strategy only in some regions.

Improving building insulation norms and climate-proofing new buildings is another typical example of no-regret strategy, since this action increases climate robustness while energy savings can often pay back the additional cost in only a few years. Considering its high cost, on the other hand, it is unlikely that the climate-proofing of existing buildings is no-regret, at least over the short term. Land-use policies that aim at limiting urbanization and development in certain flood-prone areas (e.g., coastal zones in Louisiana or Florida) would reduce disaster losses in the present climate, and climate change may only make them more desirable. Also, in many locations, especially coastal cities, building sea walls would be economically justified by storm surge risks with the current sea level (Hanson et al. 2011), and sea level rise will only make these walls more socially beneficial.

**The idea is therefore not to design adaptation strategies assuming that the present situation is optimal and should be preserved in spite of climate change. Instead, the identification of sub-optimalities in the current situation may help identify adaptation options that are beneficial over the short term (i.e., easier to implement from a political point of view) and efficient enough to reduce long-term climate vulnerability.**

**When considering no regret strategies, it is critical to investigate why they have not been already implemented. Many obstacles explain the current situation, including (i) financial and technology constraints, especially in poor countries; (ii) lack of information and transaction costs at the micro-level; and (iii) institutional and legal constraints.**

**Table 7.2** Examples of adaptation options in various sectors, and their assessment in light of the strategies proposed by this article

| Sector | Examples of adaptation options | No regret strategy | Reversible/ flexible | Existence of cheap safety margins | Soft strategy | Reduced decision horizon | Synergies with mitigation | Ranking |
|---|---|---|---|---|---|---|---|---|
| Agriculture | Developing crop insurance | + | + | | + | | | 1 |
| | Irrigation (possibly with water storage and transport) | + | – | + | | | – | 2 |
| | Forestry with shorter rotation time | – | – | | | + | – | 2 |
| | Development of resistant crops | + + | | | | | | 1 |
| Coastal zones | Coastal defenses/sea walls | + | – | + | | | | 2 |
| | "Easy-to-retrofit" defenses | | + | + | | + | | 1 |
| | Enhanced drainage systems | + | – | + | | | | 2 |
| | Restrictive land use planning | + | + | + | | | | 1 |
| | Insurance, warning and evacuation schemes | + + | + | | + | | | 1 |
| | Relocation and retreat | – | – | | | | | 3 |
| | Creation of risk analysis institution and long-term plans | + | + | | + | | | 1 |
| Health and housing | Air conditioning | | | | | | – | 3 |
| | Improved building standards | + | | + | | | + | 1 |
| | R&D on vector control, vaccines | + | | | | | | 1 |
| | Improvements in public health systems | + + | | | | | | 1 |

(continued)

**Table 7.2** (continued)

| Sector | Examples of adaptation options | No regret strategy | Reversible/ flexible | Existence of cheap safety margins | Soft strategy | Reduced decision horizon | Synergies with mitigation | Ranking |
|---|---|---|---|---|---|---|---|---|
| Water resources | Institutionalization of long-term prospective | ++ | + | | + | | | 1 |
| | Loss reduction (leakage control, etc.) | ++ | | | | | | 1 |
| | Demand control and water reuse | ++ | + | | + | | | 1 |
| | Storage capacity increase (new reservoirs) | + | − | + | | | | 2 |
| | Desalination and water transport | + | − | + | | | − | 3 |
| Human settlements | Climate proofing of new building and infrastructure | + | | + | | | + | 1 |
| | Climate proofing of old building and infrastructure | | − | | | | + | 2 |
| | Improvement of urban infrastructures | + | − | + | | | + | 2 |
| | Restrictive land use planning | ++ | + | + | | | | 1 |
| | Flood barriers, storm/flood proof infrastructure | + | − | + | | | | 2 |
| | Development of early warning systems | ++ | + | | + | | | 1 |

In the "no regret" column, "++" indicates options that yield benefits even with no climate change in most cases, while "+" indicates options that are no-regret only in some cases, depending on local characteristics. The last column provides a three-category ranking of the options that should be favored, based on this analysis

**The investigation of why these measures have not been implemented yet will suggest how they can be implemented tomorrow.** If the constraint is related to the financing of the measure (i.e. it is cost-efficient but requires high up-front costs that are difficult to mobilize), then the creation of a fund can be a solution. If the constraint is related to institutional constraints, there is no need for a fund and more resources, but a policy reform is needed.

### *7.2.2 Reversible Strategies*

**In an uncertain context, and when no-regret strategies are not available, it is wise to favor strategies that are reversible and flexible over irreversible choices. The aim is to keep as low as possible the cost of being wrong about future climate change.** Among these examples, one can mention "easy-to-retrofit" defenses; i.e., defenses initially designed to allow for cheap upgrades if sea level rise makes them insufficient; the climate proofing of new buildings and infrastructure, which has an immediate cost but can be stopped instantaneously if new information shows that this measure is finally unnecessary; and insurance and early warning systems that can be adjusted every year in response to the arrival of new information.

Another example is restrictive urban planning. When deciding whether to allow the urbanization of an area potentially at risk of flooding if climate change increases river runoff, the decision-maker must be aware of the fact that one answer is reversible while the other is not. Refusing to urbanize, indeed, has a well-known short-term cost, but if new information shows in the future that the area is safe, urbanization can be allowed virtually overnight. This option, therefore, is highly reversible, even though it is not costless since it may prevent profitable investments from being realized. Allowing urbanization now, on the other hand, yields short-term benefits, but if the area is found dangerous in the future, the choice will be between retreat and protection. Retreat is very difficult politically, especially if urbanization has been explicitly allowed. Protection is also expensive, and it is important to consider the residual risk: protection is efficient up to the protection design. If the protection is overtopped or fails, human and economic losses can be very large. So, allowing urbanization is very difficult to reverse, and this strategy is highly vulnerable to the underestimation of future risks. Of course, it does not mean that urbanization should always be rejected. It only means that, in the decision-making process, the value of the reversibility of a strategy, often referred to as the "option value," should be taken into account.

The option value is often used to assess the possibility of delaying a decision (Ha-Duong 1998), as in this urbanization example. For many infrastructure decisions, however, waiting is not an option, since all climate-sensitive decisions (e.g., in water management or housing) cannot simply be delayed by decades. The valuation of reversibility, through the option value concept or through multi-criteria decision-making frameworks, have thus to be applied to the comparison of adaptation strategies with different "irreversibility levels."

### 7.2.3 Safety-Margin Strategies

**When it is impossible to implement a flexible strategy, it is useful to introduce low-cost "safety margins" in project designs.**

The existence of such strategies to manage sea level rise or water investments has been mentioned by Nicholls and Leatherman (1996), Groves and Lempert (2007), and Groves et al. (2007). And there are practical applications today. For instance, to calibrate drainage infrastructure, water managers in Copenhagen now use runoff figures that are 70 % larger than their current level. Some of this increase is meant to deal with population growth and the rest is to cope with climate change, which may lead to an increase in heavy precipitation over Denmark. This 70 % increase has not been precisely calibrated, because such a calibration is made impossible by climate change uncertainty. But this increase is thought to be large enough to cope with almost any possible climate change during this century, considering the information provided by all climate models. This move is justified by the fact that, in the design phase, it is inexpensive to implement a drainage system able to cope with increased precipitation. On the other hand, modifying the system after it has been built is difficult and expensive. It is wise, therefore, to be over-pessimistic in the design phase.

The same is often true for dikes and sea walls: construction costs alone are often manageable (see, e.g., The Foresight report on Flood and Coastal Defences, Volume 2, Table 5.2., available on http://www.foresight.gov.uk); a significant fraction of the total social cost of a dike arising from amenity costs (e.g., loss of sea view), and other indirect effects (e.g., loss of biodiversity, other environmental costs on ecosystems, or enhanced erosion in neighboring locations). As a consequence, the marginal cost to build a higher dam is small compared to its total cost. If a dike has to be built today to cope with current storm surge risks, therefore, it may be justified to build it higher, in such a way that it can cope with future sea levels.

Often, when it is cheap, it is sensible to add "security margins" to design criteria, in order to improve the resilience of infrastructure to future (expected or unexpected) changes. Cheap safety margins can be introduced in many existing adaptation options, to take into account climate uncertainty: developing drainage infrastructures in developing country cities can be considered as an adaptation measure; making these drainage infrastructures able to cope with more water than we currently expect is a "safety-margin" strategy that makes this adaptation measure more robust.

The existence of cheap safety margins is especially important for adaptation measures that are not reversible or flexible. The options that are irreversible (e.g., retreat from coastal areas) and in which no cheap safety margins are available are particularly inadequate in the current context. The options that are irreversible but in which safety margins can be introduced (e.g., coastal defenses or improvement of urban water-management infrastructures) can be implemented, but only with a careful taking into account of future climate change scenarios.

### 7.2.4 *Soft Strategies*

Technical solutions are not the only way of adapting to changing climates. Sometimes, institutional or financial tools can also be efficient. For instance, the institutionalization of a long-term planning horizon may help anticipate problems and implement adequate responses: in the framework of the California Water Plan, all water suppliers that provide water to more than 3,000 customers in California have to carry out, every 5 years, a 25-year prospective of their activity, including the anticipation of future water demand, future water supply sources, and "worst-case" drought scenarios. These kinds of exercises are very useful because they force planners to think several decades ahead, they create contacts between economic agents and climate scientists, and they help shape strategies to cope with future changes. In the present situation, where parameters that used to be known become uncertain, a long-term planning horizon is key to determining where and how to change business practices.

Institutional solutions have also an important role to play in coastal zone management: while managing coastal floods did not require regular updates in a world with an almost constant sea level, climate change and sea level rise will make it necessary to analyze coastal flood risks on a regular basis and to implement upgrades when required. The creation of specific institutions to carry out these analyses may, therefore, be an efficient adaptation option.

In the same way, in hurricane-prone regions, it may be more efficient to implement an efficient warning and evacuation system combined with a strong (possibly expensive) insurance scheme and recovery plan than to protect all populations with seawalls and dikes. In the former case, the population is evacuated in dangerous conditions (e.g., an approaching hurricane) to avoid deaths and casualties, and material losses are paid by insurance claims, so that recovery and reconstruction are as effective as possible. The insurance premium the population will have to pay to live in this at-risk area may be large, but remains lower than the cost of protecting the areas with dikes. Of course, warning systems are not flawless and it is always difficult to decide whether and when to evacuate, but the Katrina experience demonstrated that hard protection can also fail, with the most tragic consequences.

Soft adaptation options are also reversible solutions. The key advantage of soft adaptation options, indeed, is that they entail much less inertia and irreversibility than hard adaptation: an insurance scheme can be adjusted every year, unlike a water reservoir. The risk of sunk costs if climate projections are wrong is much lower for institutional and financial strategies than for technical adaptation projects, which makes them more suitable to the current context of high uncertainty.

An important caveat is that soft options like land-use plans, insurance schemes, and early warning systems will have an influence on business investment choices and household decisions and, therefore, on hard investments. For instance, land-use planning restrictions can be seen as soft options, but their consequences in terms of construction make such a qualification questionable. As a consequence, no option is purely a soft option.

### *7.2.5 Strategies That Reduce Decision-Making Time Horizons*

The uncertainty regarding future climate conditions increases rapidly with decision time horizons. Reducing the lifetime of investments, therefore, is an option to reduce uncertainty and corresponding costs. This strategy has already been implemented in the forestry sector by choosing species that have a shorter rotation time. Since species choice cannot be made reversible and no safety margins are available in this sector, this option is interesting in spite of its cost. Sometimes, it may even be preferable to shift from forestry to annual crops, for which long term uncertainty does not matter.

In other sectors, it is also often possible to avoid long-term commitment and choose shorter-lived decisions. For example, if houses will be built in an area that may become at risk of flooding if precipitation increases, it may be rational to build cheaper houses with a shorter lifetime instead of high-quality houses meant to last 100 years.

### *7.2.6 Taking into Account Conflicts and Synergies*

A last point deserves to be mentioned. Adaptation strategies often have side effects that can be either negative or positive. For instance, in the case of coastal infrastructure to protect against storm surge such as sea walls, these may threaten the tourism industry because they change landscape, ecosystem health, and beach leisure attractions. Coastal attractiveness for leisure and tourism activities is closely linked to various parameters such as landscapes (Lothian 2006), the quality of the environment, and water availability. As a consequence, in some contexts, hard protection would simply not be an option. Equally important, hard protection could contribute to damaging coastal ecosystems.

There are also conflicts between adaptation options. For instance, an increased use of snow-making to compensate for shorter skiing seasons in mountain areas would have negative consequences for water availability and – for example – agriculture. This example shows that adaptation strategies that look profitable when considering only one sector may be suboptimal at the macroeconomic scale because of negative externalities. As a consequence, public authorities will have to be aware of this risk and monitor the emergence of new externalities from adaptation behaviors.

Adaptation also interacts with mitigation policies. For example, improved building norms would lead to large ancillary benefits in terms of energy consumption and reduced greenhouse gas emissions. And indeed, the benefits in terms of emission reduction of several adaptation options can make these measures interesting, even when they imply some irreversibility. But conflicts may also appear between adaptation and mitigation measures. Many adaptation strategies that are appealing today imply increased energy consumption, like a generalization of air conditioning.

In the design of adaptation strategies, therefore, future energy costs have to be taken into account: if there is a high carbon price in 2030, desalinization plants using fossil fuels may become excessively expensive to run. Considering the huge investment cost of these plants, this possibility has to be accounted for in the decision-making process. Moreover, there is an unfortunate correlation between energy costs and climate change impacts. If climate change and its impacts appear to be worse than expected in 50 years, stricter mitigation strategies are likely to be introduced, making energy costs and carbon prices rise. Highly energy-consuming adaptation options, therefore, seem to be particularly non-robust to unexpected climate-related changes.

Finally, there are conflicts between adaptation strategies and other policy goals, and no strategy can be implemented if these conflicts are not acknowledged. Building norms can be modified to make buildings more resilient to heat waves, but this would raise construction costs, which may be a problem in countries or regions with housing scarcity (e.g., Paris and its region). Also, different building norms, and building retrofitting for higher temperatures, would modify the external aspects of buildings and cities. This move could therefore be opposed on the ground of patrimonial protection: does the population want to keep an historical neighborhood as it is, or to change it to improve comfort and living conditions? Solving these debates often requires going beyond a top-down approach in which adaptation strategies are developed by experts on the basis of scientific information. Participatory approaches, in particular, help identify which strategies are consistent with the local context and goals, and select no-regret strategies that answer other demands from the population.

## References


Dessai S, Hulme M, Lempert R, Pielke R Jr (2009) Climate prediction: a limit to adaptation? In: Adger WN, Lorenzoni I, O'Brien KL (eds) Adapting to climate change: thresholds, values, governance. Cambridge University Press, Cambridge, pp 64–78

Fankhauser S (1994) Protection vs. retreat – the economic costs of sea level rise. Environ Plan A 27:299–319

Fankhauser S, Smith JB, Tol RSJ (1999) Weathering climate change: some simple rules to guide adaptation decisions. Ecol Econ 30(1):67–78

Groves DG, Lempert RJ (2007) A new analytic method for finding policy-relevant scenarios. Glob Environ Chang 17:73–85

Groves DG, Knopman D, Lempert R, Berry S, Wainfan L (2007) Presenting uncertainty about climate change to water resource managers—summary of workshops with the inland empire utilities agency. RAND, Santa Monica

Ha-Duong M (1998) Quasi-option value and climate policy choices. Energy Econ 20:599–620

Hall JW (2007) Probabilistic climate scenarios may misrepresent uncertainty and lead to bad adaptation decisions. Hydrol Process 21(8):1127–1129

Hallegatte S (2009) Strategies to adapt to an uncertain climate change. Glob Environ Chang 19:240–247

Hallegatte S, Shah A, Brown C, Lempert R, Gill S (2012) Investment decision making under deep uncertainty–application to climate change. World Bank Policy Research working paper 6193. Washington, DC, USA

Hanson S, Nicholls R, Ranger N, Hallegatte S, Corfee-Morlot J, Herweijer C, Chateau J (2011) A global ranking of port cities with high exposure to climate extremes. Clim Chang 104(1):89–111

Lempert RJ, Collins MT (2007) Managing the risk of uncertain thresholds responses: comparison of robust, optimum, and precautionary approaches. Risk Anal 27:1009–1026

Lempert RJ, Groves DG, Popper SW, Bankes SC (2006) A general, analytic method for generating robust strategies and narrative scenarios. Manag Sci 52(4):514–528

Lempert RJ, Kalra N, Peyraud S, Mao Z (2013) Ensuring robust flood risk management in Ho Chi Minh City: a robust decision making demonstration. World Bank Policy Research working paper 6465. Washington, DC, USA

Lothian A (2006) Coastal landscape assessment. In: Coast to coast conference, Melbourne, 23 May 2006

Milly PCD, Betancourt J, Falkenmark M, Hirsch RM, Kundzewicz ZW, Lettenmaier DP, Stouffer RJ (2008) Climate change: stationarity is dead: whither water management? Science 319(5863):573

Nicholls R, Leatherman S (1996) Adapting to sea-level rise: relative sea-level trends to 2100 for the United States. Coast Manag 24(4):301–324

Schwartz P (1996) The art of the long view. Double Day, New York

CPSIA information can be obtained at www.ICGtesting.com
Printed in the USA
BVOW11*2340140115

383413BV00001B/2/P

9 783319 089324